CALCULUS

SECOND EDITION

Finney/Thomas

Student's Solutions Manual
Special Edition Part II

Michael B. Schneider
Thomas L. Cochran

Belleville Area College

ADDISON-WESLEY

An imprint of Addison Wesley Longman, Inc.

Reading, Massachusetts • Menlo Park, California • New York • Harlow, England
Don Mills, Ontario • Sydney • Mexico City • Madrid • Amsterdam

Reproduced by Addison-Wesley from camera-ready copy supplied by the author.

ISBN 0-201-31129-1
1 2 3 4 5 6 7 8 9 10-CRC-00999897

TABLE OF CONTENTS

CHAPTER 9

INFINITE SERIES

9.1 SEQUENCES

1. $a_1 = 0, a_2 = -\frac{1}{4}, a_3 = -\frac{2}{9}, a_4 = -\frac{3}{16}$

5. $(-1)^2, (-1)^3, (-1)^4, (-1)^5, \ldots \Rightarrow a_n = (-1)^{n+1}$

9. $1^2 - 1, 2^2 - 1, 3^2 - 1, 4^2 - 1, \ldots \Rightarrow a_n = n^2 - 1$

13. $1, \frac{3}{2}, \frac{7}{4}, \frac{15}{8}, \frac{31}{16}, \frac{63}{32}, \frac{127}{64}, \frac{255}{128}, \frac{511}{256}, \frac{1023}{512}$

17. $(1), (1), (1 + 1), (1 + 1) + (1), (1 + 1 + 1) + (1 + 1), (1 + 1 + 1 + 1 + 1) + (1 + 1 + 1) \ldots = 1, 1, 2, 3, 5, 8,$ $13, 21, 34, 55$

21. $\lim_{n\to\infty} a_n = \lim_{n\to\infty} \frac{1 - 2n}{1 + 2n} = -1 \Rightarrow$ converges

25. $\lim_{n\to\infty} a_n = \lim_{n\to\infty} \frac{n^2 - 2n + 1}{n - 1} = \lim_{n\to\infty} n - 1 = \infty \Rightarrow$ diverges

29. $\lim_{n\to\infty} a_n = \lim_{n\to\infty} \left(\frac{n+1}{2n}\right)\left(1 - \frac{1}{n}\right) = \frac{1}{2} \Rightarrow$ converges

33. $\lim_{n\to\infty} a_n = \lim_{n\to\infty} \frac{\sin n}{n} = 0$, by the Sandwich Theorem for Sequences $\Rightarrow$ converges

37. $\lim_{n\to\infty} a_n = \lim_{n\to\infty} \tan^{-1} n = \frac{\pi}{2} \Rightarrow$ converges

41. $\lim_{n\to\infty} a_n = \lim_{n\to\infty} \frac{\ln(n+1)}{\sqrt{n}} = 0$, due to the growth rates in section 6.7 $\Rightarrow$ converges

45. $\lim_{n\to\infty} a_n = \lim_{n\to\infty} \left(1 + \frac{7}{n}\right)^n = e^7$, due to table 9.1 part 5 $\Rightarrow$ converges

49. $\lim_{n\to\infty} a_n = \lim_{n\to\infty} \sqrt[n]{10n} = \lim_{n\to\infty} \exp\left(\frac{\ln(10n)}{n}\right) = \exp\left(\lim_{n\to\infty} \frac{\ln 10n}{n}\right) = \exp\left(\lim_{n\to\infty} \frac{1}{n}\right) = e^0 = 1 \Rightarrow$ converges; $\exp[f(n)] = e^{f(n)}$.

53. $\lim_{n\to\infty} a_n = \lim_{n\to\infty} \frac{\ln n}{n^{1/n}} = \frac{\lim_{n\to\infty} \ln n}{\lim_{n\to\infty} \sqrt[n]{n}} = \frac{\infty}{1} = \infty$, due to table 9.1, part 2 $\Rightarrow$ diverges

57. $\lim_{n\to\infty} a_n = \lim_{n\to\infty} \frac{n!}{n^n} = \lim_{n\to\infty} \frac{1 \cdot 2 \cdot 3 \cdots (n-1)(n)}{n \cdot n \cdot n \cdots n \cdot n} \le \lim_{n\to\infty} \frac{1}{n} = 0$ and $\frac{n!}{n^n} \ge 0 \Rightarrow \lim_{n\to\infty} \frac{n!}{n^n} = 0 \Rightarrow$ converges

61. $\lim_{n\to\infty} a_n = \lim_{n\to\infty} \frac{n!}{10^{6n}} = \frac{1}{\lim_{n\to\infty} \frac{(10^6)^n}{n!}} = \infty$, due to table 9.1, part 6 $\Rightarrow$ diverges

65. $\lim_{n\to\infty} a_n = \lim_{x\to\infty} \ln\left(1+\frac{1}{n}\right)^n = \ln\left(\lim_{n\to\infty}\left(1+\frac{1}{n}\right)^n\right) = \ln e = 1 \Rightarrow$ converges

69. $\lim_{n\to\infty} a_n = \lim_{n\to\infty} \frac{n^2\sin(1/n)}{2n-1} = \lim_{n\to\infty} \frac{\sin\left(\frac{1}{n}\right)}{\frac{2}{n}-\frac{1}{n^2}} = \lim_{n\to\infty} \frac{-\left(\cos\left(\frac{1}{n}\right)\right)\left(\frac{1}{n^2}\right)}{-\frac{2}{n^2}+\frac{2}{n^3}} = \lim_{n\to\infty} \frac{-\cos(1/n)}{-2+2/n} = \frac{1}{2} \Rightarrow$ converges

73. $\left|\sqrt[n]{0.5}-1\right| < 10^{-3} \Rightarrow -\frac{1}{1000} < \left(\frac{1}{2}\right)^{1/n} - 1 < \frac{1}{1000} \Rightarrow n > \frac{\ln(1/2)}{\ln(999/1000)} \Rightarrow n > 692.8 \Rightarrow N = 692$

77. a) No, $\lim_{n\to\infty} A_n = \lim_{n\to\infty} A_o\left(1+\frac{r}{12}\right)^n = A_o \lim_{n\to\infty}\left(\frac{12+r}{12}\right)^n = \infty$, since $\frac{12+r}{12} > 1$

b) $\{A_n\} = \left\{(\$\,1000)\left(1+\frac{0.02583^n}{12}\right)\right\} = \left\{(\$\,1000)(1.0021525)^n\right\} \Rightarrow A_1 = \$\,1002.15$, $A_2 = \$\,1004.31$, $A_3 = \$\,1006.47$, $A_4 = \$\,1008.64$, $A_5 = \$\,1010.81$, $A_6 = \$\,1012.98$, $A_7 = \$\,1015.17$, $A_8 = \$\,1017.35$, $A_9 = \$\,1019.54$, $A_{10} = \$\,1021.73$

81. a) $f(x) = x^2 - 2$; the sequence converges to $1.414213562 \approx \sqrt{2}$

b) $f(x) = \tan(x) - 1$; the sequence converges to $0.7853981635 \approx \frac{\pi}{4}$

c) $f(x) = e^x$; the sequence $1, 0, -1, -2, -3, -4, -5 \cdots$ diverges

9.2 INFINITE SERIES

1. $s_n = \frac{a(1-r^n)}{(1-r)} = \frac{2(1-(1/3)^n)}{1-1/3} \Rightarrow \lim_{n\to\infty} s_n = \frac{2}{1-1/3} = 3$

5. $\frac{1}{(n+1)(n+2)} = \frac{1}{n+1} - \frac{1}{n+2} \Rightarrow s_n = \left(\frac{1}{2}-\frac{1}{3}\right) + \left(\frac{1}{3}-\frac{1}{4}\right) + \ldots + \left(\frac{1}{n+1}-\frac{1}{n+2}\right) = \frac{1}{2} - \frac{1}{n+2} \Rightarrow \lim_{n\to\infty} s_n = \frac{1}{2}$

9. $\frac{7}{4} + \frac{7}{16} + \frac{7}{64} + \ldots$, the sum of this geometric series is $\frac{7/4}{1-1/4} = \frac{7}{3}$

13. $(1+1) + \left(\frac{1}{2}-\frac{1}{5}\right) + \left(\frac{1}{4}+\frac{1}{25}\right) + \left(\frac{1}{8}-\frac{1}{125}\right) + \ldots$, is the sum of two geometric series; the sum $\frac{1}{1-1/2} + \frac{1}{1+1/5} = 2 + \frac{5}{6} = \frac{17}{6}$

$n \to \infty \qquad n \to \infty$

17. $\dfrac{40n}{(2n-1)^2(2n+1)^2} = \dfrac{A}{(2n-1)} + \dfrac{B}{(2n-1)^2} + \dfrac{C}{(2n+1)} + \dfrac{D}{(2n+1)^2} =$

$\dfrac{A(2n-1)(2n+1)^2 + B(2n+1)^2 + C(2n+1)(2n-1)^2 + D(2n-1)^2}{(2n-1)^2(2n+1)^2} = \Rightarrow$

$A(2n-1)(2n+1)^2 + B(2n+1)^2 + C(2n+1)(2n-1)^2 + D(2n-1)^2 = 40n \Rightarrow A\left(8n^3 + 4n^2 - 2n - 1\right) + B\left(4n^2 + 4n + 1\right) + C\left(8n^3 - 4n^2 - 2n + 1\right) + D\left(4n^2 - 4n + 1\right) = 40n \Rightarrow (8A + 8C)n^3 + (4A + 4B - 4C + 4D)n^2 + (-2A + 4B - 2C - 4D)n + (-A + B + C + D) = 40n \Rightarrow \begin{cases} 8A + 8C = 0 \\ 4A + 4B - 4C + 4D = 0 \\ -2A + 4B - 2C - 4D = 40 \\ -A + B + C + D = 0 \end{cases} \Rightarrow \begin{cases} 8A + 8C = 0 \\ A + B - C + D = 0 \\ -A + 2B - C - 2D = 20 \\ -A + B + C + D = 0 \end{cases} \Rightarrow$

$\begin{cases} B + D = 0 \\ 2B - 2D = 20 \end{cases} \Rightarrow 4B = 20 \Rightarrow B = 5 \text{ and } D = -5 \Rightarrow \begin{cases} A + C = 0 \\ -A + 5 + C - 5 = 0 \end{cases} \Rightarrow C = 0 \text{ and } A = 0.$ Hence,

$$\sum_{n=1}^{k} \left(\frac{40n}{(2n-1)^2(2n+1)^2}\right) = 5 \sum_{n=1}^{k} \left(\frac{1}{(2n-1)^2} - \frac{1}{(2n+1)^2}\right) =$$

$5\left(\dfrac{1}{1} - \dfrac{1}{9} + \dfrac{1}{9} - \dfrac{1}{25} + \dfrac{1}{25} - \ldots - \dfrac{1}{\left(2(n-1)+1\right)^2} + \dfrac{1}{(2n-1)^2} - \dfrac{1}{(2n+1)^2}\right) = 5\left(1 - \dfrac{1}{(2n+1)^2}\right).$ Therefore, the sum is

$\lim\limits_{n \to \infty} 5\left(1 - \dfrac{1}{(2n+1)^2}\right) = 5.$

21. A convergent geometric series with a sum of $\dfrac{3/2}{1 - (-1/2)} = 1$

25. A convergent geometric series with a sum of $\dfrac{1}{1 - 1/e^2} = \dfrac{e^2}{e^2 - 1}.$

29. The difference of two convergent geometric series with a sum of $\dfrac{1}{1 - 2/3} - \dfrac{1}{1 - 1/3} = 3 - \dfrac{3}{2} = \dfrac{3}{2}.$

33. $\displaystyle\sum_{n=1}^{\infty} \ln\left(\frac{n}{n+1}\right) = \sum_{n=1}^{\infty} \ln(n) - \ln(n+1) \Rightarrow s_n = \left(\ln(1) - \ln(2)\right) + \left(\ln(2) - \ln(3)\right) + \left(\ln(3) - \ln(4)\right) + \ldots + \left(\ln(n-1) - \ln(n)\right) + \left(\ln(n) - \ln(n+1)\right) = \ln(1) - \ln(n+1) = -\ln(n+1) \Rightarrow \lim_{n \to \infty} s_n = -\infty, \Rightarrow$ the series diverges

37. $\displaystyle\sum_{n=1}^{\infty} \left(\frac{1}{\ln(n+2)} - \frac{1}{\ln(n+1)}\right) \Rightarrow s_n = \left(\frac{1}{\ln 3} - \frac{1}{\ln 2}\right) + \left(\frac{1}{\ln 4} - \frac{1}{\ln 3}\right) + \left(\frac{1}{\ln 5} - \frac{1}{\ln 4}\right) + \ldots + \left(\frac{1}{\ln(n+1)} - \frac{1}{\ln(n)}\right) + \left(\frac{1}{\ln(n+2)} - \frac{1}{\ln(n+1)}\right) = -\frac{1}{\ln 2} + \frac{1}{\ln(n+2)} \Rightarrow \lim_{n \to \infty} s_n = \lim_{n \to \infty} \left(-\frac{1}{\ln 2} + \frac{1}{\ln(n+2)}\right) = -\frac{1}{\ln 2} \Rightarrow$ the series converges

41. $a = 3,\ r = \dfrac{x-1}{2}$ where $\left|\dfrac{x-1}{2}\right| < 1 \Rightarrow -1 < x < 3$

45. $0.\overline{7} = \dfrac{7}{10} + \dfrac{7}{10}\left(\dfrac{1}{10}\right) + \dfrac{7}{10}\left(\dfrac{1}{10}\right)^2 + \cdots = \dfrac{7}{9}$

49. $1.24\overline{123} = \frac{31}{25} + \frac{123}{10^5} + \frac{123}{10^5}\left(\frac{1}{10^3}\right) + \frac{123}{10^5}\left(\frac{1}{10^3}\right)^2 + \cdots = \frac{31}{25} + \frac{123}{99900} = \frac{41333}{33300}$

53. a) $\sum_{n=-2}^{\infty} \frac{1}{(n+4)(n+5)}$ b) $\sum_{n=0}^{\infty} \frac{1}{(n+2)(n+3)}$ c) $\sum_{n=5}^{\infty} \frac{1}{(n-3)(n-2)}$

57. a) $L_1 = 3, L_2 = 3\left(\frac{4}{3}\right), L_3 = 3\left(\frac{4}{3}\right)^2, \ldots, L_n = 3\left(\frac{4}{3}\right)^{n-1}$. $\therefore \lim_{n \to \infty} L_n = \lim_{n \to \infty} 3\left(\frac{4}{3}\right)^{n-1} = \infty$

b) $A_1 = \frac{1}{2}(1)\left(\frac{\sqrt{3}}{2}\right) = \frac{\sqrt{3}}{4}$, $A_2 = A_1 + 3\left(\frac{1}{2}\right)\left(\frac{1}{3}\right)\left(\frac{\sqrt{3}}{6}\right) = \frac{\sqrt{3}}{4} + \frac{\sqrt{3}}{12}$, $A_3 = A_1 + A_2 + 12\left(\frac{1}{2}\right)\left(\frac{1}{9}\right)\left(\frac{\sqrt{3}}{18}\right) = \frac{\sqrt{3}}{4} + \frac{\sqrt{3}}{12} + \frac{\sqrt{3}}{27}$, $A_4 = A_1 + A_2 + 48\left(\frac{1}{2}\right)\left(\frac{1}{27}\right)\left(\frac{\sqrt{3}}{54}\right), \ldots, A_n = \frac{\sqrt{3}}{4} + \frac{27\sqrt{3}}{64}\left(\frac{4}{9}\right)^2 + \frac{27\sqrt{3}}{64}\left(\frac{4}{9}\right)^3 + \cdots = \frac{\sqrt{3}}{4} + \sum_{n=2}^{\infty} \frac{27\sqrt{3}}{64}\left(\frac{4}{9}\right)^n = \frac{2\sqrt{3}}{5}$

61. Let $a_n = (1/4)^n$ and $b_n = (1/2)^n$. Then $A = \sum_{n=1}^{\infty} a_n = 1/3$, $B = \sum_{n=1}^{\infty} b_n = 1$ and $\sum_{n=1}^{\infty} (a_n / b_n) = \sum_{n=1}^{\infty} (1/2)^n = 1 \neq A/B$.

9.3 SERIES WITH NONNEGATIVE TERMS–COMPARISON AND INTEGRAL TESTS

1. converges, a geometric series with $r = 1/10$

5. diverges, by the Limit Comparison Test when compared with $\sum_{n=1}^{\infty} \frac{1}{\sqrt{n}}$ a divergent p–series

9. converges, a geometric series with $r = 1/8$

13. converges, a geometric series with $r = 2/3$

17. diverges, $\lim_{n \to \infty} a_n = \lim_{n \to \infty} \frac{2^n}{n+1} = \lim_{n \to \infty} \frac{2^n \ln 2}{1} = \infty \neq 0$

21. diverges, by the Direct Comparison Test for Convergence, since $\frac{1}{n} \le \frac{\sqrt{n}}{n} \le \frac{\sqrt{n}}{\ln n}$ for $n \ge 2$

25. converges, by the Limit Comparison Test when compared with $\sum_{n=1}^{\infty} \frac{1}{n^2}$, a convergent p–series, for

$$\lim_{n \to \infty} \frac{(\ln n)^2 / n^3}{1/n^2} = \lim_{n \to \infty} \frac{(\ln n)^2}{n} = \lim_{n \to \infty} \frac{2(\ln n)(1/n)}{1} = 2 \lim_{n \to \infty} \frac{\ln n}{n} = 2 \lim_{n \to \infty} \frac{1/n}{1} = 2 \cdot 0 = 0$$

29. diverges, $\lim_{n\to\infty} a_n = \lim_{n\to\infty} \left(1+\frac{1}{n}\right)^n = e \neq 0$

33. diverges, a geometric series with $r = 1/\ln 2 \approx 1.44 > 1$

37. diverges, by the Limit Comparison Test when compared to the divergent harmonic series since $\lim_{n\to\infty} \frac{\sin(1/n)}{1/n} = \lim_{x\to 0} \frac{\sin x}{x} = 1 \neq 0$

41. converges, by the Integral Test for $\int_1^\infty \frac{\tan^{-1} x}{1+x^2}\,dx = \int_{\pi/4}^{\pi/2} u\,dx = u = \left[\frac{u^2}{2}\right]_{\pi/4}^{\pi/2} = \frac{1}{2}\left[(\pi/2)^2 - (\pi/4)^2\right] = \frac{1}{2}\left[\frac{\pi^2}{4} - \frac{\pi^2}{16}\right] = \frac{\pi^2}{8} - \frac{\pi^2}{32} = \frac{4\pi^2 - \pi^2}{32} = \frac{3\pi^2}{32}$ where $u = \tan^{-1} x$, $du = \frac{1}{1+x^2}\,dx$, the lower limit $u = \tan^{-1} 1 = \pi/4$ and the upper limit $u = \lim_{x\to\infty} \tan^{-1} x = \pi/2$

45. converges, by the Direct Comparison Test for Convergence, since $\frac{\tan^{-1} n}{n^{1.1}} < \frac{\pi/2}{n^{1.1}}$ and $\sum_{n=1}^{\infty} \frac{\pi/2}{n^{1.1}} = \frac{\pi}{2} \sum_{n=1}^{\infty} \frac{1}{n^{1.1}}$ the product of a convergent p–series and a nonzero constant

49. converges, by the Limit Comparison Test when compared with $\sum_{n=1}^{\infty} \frac{1}{e^n}$, a convergent geometric series, for

$$\lim_{n\to\infty} \frac{2/\left(1+e^n\right)}{1/e^n} = \lim_{n\to\infty} \frac{2e^n}{1+e^n} = \lim_{n\to\infty} \frac{2e^n}{e^n} = \lim_{n\to\infty} 2 = 2$$

53. converges, by the Limit Comparison Test when compared with $\sum_{n=1}^{\infty} \frac{1}{n^2}$, a convergent p–series, for

$$\lim_{n\to\infty} \frac{(\coth n)/n^2}{1/n^2} = \lim_{n\to\infty} \coth n = \lim_{n\to\infty} \frac{e^n + e^{-n}}{e^n - e^{-n}} = \lim_{n\to\infty} \frac{1 + 1/e^{2n}}{1 - 1/e^{2n}} = 1$$

57. Yes, If $\sum_{n=1}^{\infty} a_n$ is a divergent series of positive numbers, then $(1/2) \sum_{n=1}^{\infty} a_n = \sum_{n=1}^{\infty} (a_n/2)$ also diverges.

There is no "smallest" divergent series of positive numbers. For any divergent series of positive numbers $\sum_{n=1}^{\infty} a_n$, $\sum_{n=1}^{\infty} (a_n/2)$ has smaller terms and still diverges.

61. a) If $p > 1$, then $\int_2^{\infty} \frac{dx}{x(\ln x)^p} dx = \lim_{b \to \infty} \int_2^b (\ln x)^{-p} \frac{1}{x} dx = \lim_{b \to \infty} \left[\frac{(\ln x)^{1-p}}{1-p}\right]_2^b =$
$\left(\frac{1}{1-p}\right) \lim_{b \to \infty} \left[(\ln b)^{1-p} - (\ln 2)^{1-p}\right] = \frac{(\ln 2)^{1-p}}{p-1}$. If $p = 1$, then $\int_2^{\infty} \frac{dx}{x \ln x} = \lim_{b \to \infty} \int_2^b \frac{1/x}{\ln x} dx =$
$\lim_{b \to \infty} \left[\ln|\ln x|\right]_2^b = \infty$. If $p < 1$, then $\int_2^{\infty} \frac{dx}{x(\ln x)^p} dx = \frac{1}{1-p} \lim_{b \to \infty} \left[(\ln b)^{1-p} - (\ln 2)^{1-p}\right] = \infty$.

b) By the Integral Test, the series would converge if $p > 1$ and diverge if $p \le 1$.

9.4 SERIES WITH NONNEGATIVE TERMS–RATIO AND ROOT TESTS

1. converges, by the Ratio Test for $\lim_{n \to \infty} \frac{a_{n+1}}{a_n} = \lim_{n \to \infty} \left|\frac{(n+1)^2}{2^{n+1}} \frac{2^n}{n^2}\right| = \frac{1}{2} < 1$

5. converges, by the Ratio Test for $\lim_{n \to \infty} \frac{a_{n+1}}{a_n} = \lim_{n \to \infty} \left|\frac{(n+1)^{10}}{10^{n+1}} \frac{10^n}{n^{10}}\right| = \frac{1}{10} < 1$

9. diverges, $\lim_{n \to \infty} a_n = \lim_{n \to \infty} \left(1 - \frac{3}{n}\right)^n = \lim_{n \to \infty} \left(1 + \frac{-3}{n}\right)^n = e^{-3} \approx 0.05 \ne 0$ (table 9.1)

13. diverges, $\sum_{n=1}^{\infty} \frac{n-1}{n^2} = \sum_{n=1}^{\infty} \frac{1}{n} - \sum_{n=1}^{\infty} \frac{1}{n^2}$, the difference between a divergent and convergent series diverges

17. converges, by the Ratio Test for $\lim_{n \to \infty} \frac{a_{n+1}}{a_n} = \lim_{n \to \infty} \left|\frac{(n+2)(n+3)}{(n+1)!} \frac{n!}{(n+1)(n+2)}\right| = 0 < 1$

21. converges, by the Ratio Test for $\lim_{n \to \infty} \frac{a_{n+1}}{a_n} = \lim_{n \to \infty} \left|\frac{1}{(2n+3)!} \frac{(2n+1)!}{1}\right| = 0 < 1$

25. converges, by the Direct Comparison Test for Convergence, since $\frac{n! \ln n}{n(n+2)!} = \frac{\ln n}{n(n+1)(n+2)} < \frac{1}{(n+1)(n+2)} <$
$\frac{1}{n^2}$ and $\sum_{n=1}^{\infty} \frac{1}{n^2}$ is a convergence p–series

29. diverges, by the Ratio Test for $\lim_{n \to \infty} \frac{a_{n+1}}{a_n} = \lim_{n \to \infty} \left|\frac{\left(\frac{3n-1}{2n+5}\right) a_n}{a_n}\right| = \frac{3}{2} > 1$, since $\frac{3n-1}{2n+5}$ is a rational function

33. converges, by the Ratio Test for $\lim_{n \to \infty} \frac{a_{n+1}}{a_n} = \lim_{n \to \infty} \frac{\frac{\ln n}{n} a_n}{a_n} = \lim_{n \to \infty} \frac{\ln n}{n} = \lim_{n \to \infty} \frac{1/n}{1} = 0 < 1$

37. converges, by the Ratio Test for $\lim_{n\to\infty} \frac{a_{n+1}}{a_n} = \lim_{n\to\infty} \left| \frac{2^{n+1}(n+1)!(n+1)!}{(2n+2)!} \cdot \frac{(2n)!}{2^n(n!)(n!)} \right| =$

$\lim_{n\to\infty} \frac{2(n+1)^2}{(2n+2)(2n+1)} = \frac{1}{2} < 1$, the limit follows since $\frac{2(n+1)^2}{(2n+2)(2n+1)}$ is a rational function

41. converges, by the Root Test for $\lim_{n\to\infty} \sqrt[n]{a_n} = \lim_{n\to\infty} \sqrt[n]{\frac{n^n}{2^{n^2}}} = \lim_{n\to\infty} \frac{n}{2^n} = \lim_{n\to\infty} \frac{1}{2^n \ln 2} = 0 < 1$

9.5 ALTERNATING SERIES AND ABSOLUTE CONVERGENCE

1. converges absolutely $\Rightarrow$ converges, by the Absolute Convergence Theorem, since $\sum_{n=1}^{\infty} |a_n|$ is $\sum_{n=1}^{\infty} \frac{1}{n^2}$ a convergent p–series

5. converges, by the Alternating Series Theorem, since $f(x) = \frac{1}{\ln x} \Rightarrow f'(x) = -\frac{1}{x(\ln x)^2} < 0 \Rightarrow f(x)$ is decreasing and hence $a_n > a_{n+1}$, $a_n > 0$ for $n \geq 1$ and $\lim_{n\to\infty} a_n = \lim_{n\to\infty} \frac{1}{\ln n} = 0$

9. converges, by the Alternating Series Theorem, since $f(x) = \frac{\sqrt{x}+1}{x+1} \Rightarrow f'(x) = \frac{1 - x - 2\sqrt{x}}{2\sqrt{x}(x+1)^2} < 0 \Rightarrow f(x)$ is decreasing and hence $a_n > a_{n+1}$, $a_n > 0$ for $n \geq 1$ and $\lim_{n\to\infty} a_n = \lim_{n\to\infty} \frac{\sqrt{n}+1}{n+1} = 0$

13. converges conditionally, since $f(x) = x^{-1/2} \Rightarrow f'(x) = \frac{-1}{2\sqrt{x^3}} < 0 \Rightarrow f(x)$ is decreasing and hence $a_n > a_{n+1}$, $a_n > 0$ for $n \geq 1$ and $\lim_{n\to\infty} \frac{1}{\sqrt{n}} = 0 \Rightarrow$ the given series converges, by the Alternating Series Test, but $\sum_{n=1}^{\infty} \frac{1}{\sqrt{n}}$ is a divergent p–series

17. converges conditionally, since $f(x) = \frac{1}{x+3} \Rightarrow f'(x) = \frac{-1}{(x+3)^2} < 0 \Rightarrow f(x)$ is decreasing and hence $a_n > a_{n+1}$, $a_n > 0$ for $n \geq 1$ and $\lim_{n\to\infty} \frac{1}{n+3} = 0 \Rightarrow$ the given series converges, by the Alternating Series Test, but $\sum_{n=1}^{\infty} \frac{1}{n+3}$ diverges, by the Direct Limit Comparison Test when compared with $\sum_{n=1}^{\infty} \frac{1}{n}$

19. diverges, by the nth–Term Test for Divergence for $\lim_{n\to\infty} a_n = \lim_{n\to\infty} \frac{3+n}{5+n} = 1 \neq 0$

23. converges absolutely, by the Absolute Convergence Theorem and the Ratio Test for $\lim_{n \to \infty} \left| \frac{a_{n+1}}{a_n} \right| =$

$$\lim_{n \to \infty} \left| \frac{(n+1)^2 \left(\frac{2}{3}\right)^{n+1}}{n^2 \left(\frac{2}{3}\right)^n} \right| = \frac{2}{3} < 1$$

27. diverges, by the nth–Term Test for Divergence since $\lim_{n \to \infty} a_n = \lim_{n \to \infty} \frac{(-1)^n n}{n+1} = \lim_{n \to \infty} (-1)^n \neq 0$

31. converges absolutely, by the Absolute Convergence Theorem and the Limit Comparison Test when $\sum_{n=1}^{\infty} \frac{1}{n^2 + 2n + 1}$ is compared with $\sum_{n=1}^{\infty} \frac{1}{n^2}$ for $\lim_{n \to \infty} \frac{1/\left(n^2 + 2n + 1\right)}{1/n^2} = \lim_{n \to \infty} \frac{n^2}{n^2 + 2n + 1} = 1$ since $\frac{n^2}{n^2 + 2n + 1}$ is a rational function

35. converges absolutely, by the Absolute Convergence Theorem and the Root Test for $\lim_{n \to \infty} \sqrt[n]{|a_n|} =$

$$\lim_{n \to \infty} \left(\frac{(n+1)^n}{(2n)^n}\right)^{1/n} = \lim_{n \to \infty} \frac{n+1}{2n} = \frac{1}{2} < 1 \text{ since } \frac{n+1}{2n} \text{ is a rational function}$$

39. converges conditionally, since $\frac{\sqrt{n+1} - \sqrt{n}}{1} \cdot \frac{\sqrt{n+1} + \sqrt{n}}{\sqrt{n+1} + \sqrt{n}} = \frac{1}{\sqrt{n+1} + \sqrt{n}}$ and $\left\{\frac{1}{\sqrt{n+1} + \sqrt{n}}\right\}$ is a decreasing sequence which converges to 0 so $\sum_{n=1}^{\infty} \frac{(-1)^n}{\sqrt{n+1} + \sqrt{n}}$ converges, but for $n > 1/3 \Rightarrow 3n > 1 \Rightarrow 4n > n+1 \Rightarrow$ $2\sqrt{n} > \sqrt{n+1} \Rightarrow 3\sqrt{n} > \sqrt{n+1} + \sqrt{n} \Rightarrow \frac{1}{3\sqrt{n}} < \frac{1}{\sqrt{n+1} + \sqrt{n}} \Rightarrow \sum_{n=1}^{\infty} \frac{1}{\sqrt{n+1} + \sqrt{n}}$ diverges by the Direct Comparison Test

43. $|\text{error}| < \left| (-1)^6 \frac{1}{5} \right| = 0.2$

47. $\frac{1}{(2n)!} < \frac{5}{10^6} \Rightarrow (2n)! > \frac{10^6}{5} = 200000 \Rightarrow 2n = 10 \Rightarrow n = 5 \Rightarrow 1 - \frac{1}{2!} + \frac{1}{4!} - \frac{1}{6!} + \frac{1}{8!} \approx$ $0.540302579 \approx 0.54030$

51. The unused terms are $\sum_{j=n+1}^{\infty} (-1)^{j+1} a_j = (-1)^{n+1}\left(a_{n+1} - a_{n+2}\right) + (-1)^{n+3}\left(a_{n+3} - a_{n+4}\right) + \ldots =$ $(-1)^{n+1}\left[\left(a_{n+1} - a_{n+2}\right) + \left(a_{n+3} - a_{n+4}\right) + \ldots\right]$. Each grouped term is positive, hence the remainder has the same sign as $(-1)^{n+1}$, which is the sign of the first unused term.

9.6 POWER SERIES

1. $\lim_{n \to \infty} \left| \frac{u_{n+1}}{u_n} \right| < 1 \Rightarrow \lim_{n \to \infty} \left| \frac{x^{n+1}}{x^n} \right| < 1 \Rightarrow -1 < x < 1$; when $x = -1$ we have $\sum_{n=1}^{\infty} (-1)^n$, a divergent series; when $x = 1$ we have $\sum_{n=1}^{\infty} 1$, a divergent series.

   a) the radius, 1; the interval of convergence, $-1 < x < 1$

   b) the interval of absolute convergence, $-1 < x < 1$

   c) there are no values for which the series is conditionally convergent

5. $\lim_{n \to \infty} \left| \frac{u_{n+1}}{u_n} \right| < 1 \Rightarrow \lim_{n \to \infty} \left| \frac{(x-2)^{n+1}}{10^{n+1}} \cdot \frac{10^n}{(x-2)^n} \right| < 1 \Rightarrow \frac{|x-2|}{10} \lim_{n \to \infty} 1 < 1 \Rightarrow \frac{|x-2|}{10} < 1 \Rightarrow |x-2| < 10 \Rightarrow$ $-10 < x - 2 < 10 \Rightarrow -8 < x < 12$; when $x = -8$ we have $\sum_{n=1}^{\infty} (-1)^n$, a divergent series; when $x = 12$ we have $\sum_{n=1}^{\infty} 1$, a divergent series.

   a) the radius, 10; the interval of convergence, $-8 < x < 12$

   b) the interval of absolute convergence, $-8 < x < 12$

   c) there are no values for which the series is conditionally convergent

9. $\lim_{n \to \infty} \left| \frac{u_{n+1}}{u_n} \right| < 1 \Rightarrow \lim_{n \to \infty} \left| \frac{x^{n+1}}{(n+1)\sqrt{n+1}\, 3^{n+1}} \cdot \frac{n\sqrt{n}\, 3^n}{x^n} \right| < 1 \Rightarrow \frac{|x|}{3}\left(\lim_{n \to \infty} \frac{n}{n+1}\right)\left(\sqrt{\lim_{n \to \infty} \frac{n}{n+1}}\right) < 1 \Rightarrow$ $\frac{|x|}{3}(1)(1) < 1 \Rightarrow |x| < 3 \Rightarrow -3 < x < 3$; when $x = -3$ we have $\sum_{i=1}^{n} \frac{(-1)^n}{n^{3/2}}$ an absolutely convergent series; when $x = 3$ we have $\sum_{i=1}^{n} \frac{1}{n^{3/2}}$ a convergent p–ser

   a) the radius, 3; the interval of convergence, $-3 \le x \le 3$

   b) the interval of absolute convergence, $-3 \le x \le 3$

   c) there are no values for which the series is conditionally convergent

13. $\lim_{n \to \infty} \left| \frac{u_{n+1}}{u_n} \right| < 1 \Rightarrow \lim_{n \to \infty} \left| \frac{x^{2n+3}}{(n+1)!} \cdot \frac{n!}{x^{2n+1}} \right| < 1 \Rightarrow x^2 \lim_{n \to \infty} \left| \frac{1}{n+1} \right| < 1$ for all x

   a) the radius, ∞, this series converges for all x

   b) this series absolutely converges for all x

   c) there are no values for which the series is conditionally convergent

15. $\lim_{n\to\infty}\left|\frac{u_{n+1}}{u_n}\right| < 1 \Rightarrow \lim_{n\to\infty}\left|\frac{x^{n+1}}{\sqrt{n^2+2n+4}}\ \frac{\sqrt{n^2+3}}{x^n}\right| < 1 \Rightarrow |x|\sqrt{\lim_{n\to\infty}\frac{n^2+3}{n^2+2n+4}} < 1 \Rightarrow$

$|x| < 1 \Rightarrow -1 < x < 1$; when $x = -1$ we have $\sum_{n=1}^{\infty}\frac{(-1)^n}{\sqrt{n^2+3}}$, a conditionally convergent series; when $x = 1$ we have $\sum_{n=1}^{\infty}\frac{1}{\sqrt{n^2+3}}$, a divergent series.

a) the radius, 1; the interval of convergence, $-1 \le x < 1$

b) the interval of absolute convergence, $-1 < x < 1$

c) This series is conditionally convergent at $x = -1$.

19. $\lim_{n\to\infty}\left|\frac{u_{n+1}}{u_n}\right| < 1 \Rightarrow \lim_{n\to\infty}\left|\frac{\sqrt{n+1}\,x^{n+1}}{3^{n+1}}\ \frac{3^n}{\sqrt{n}\,x^n}\right| < 1 \Rightarrow \frac{|x|}{3}\sqrt{\lim_{n\to\infty}\frac{n+1}{n}} < 1 \Rightarrow \frac{|x|}{3}(1) < 1 \Rightarrow |x| < 3 \Rightarrow$

$-3 < x < 3$; both series $\sum_{n=1}^{\infty}\sqrt{n}\,(-1)^n$ and $\sum_{n=1}^{\infty}\sqrt{n}$ diverges, when $x = \pm 3$.

a) the radius, 3; the interval of convergence, $-3 < x < 3$

b) the interval of absolute convergence, $-3 < x < 3$

c) there are no values for which the series is conditionally convergent

23. $\lim_{n\to\infty}\left|\frac{u_{n+1}}{u_n}\right| < 1 \Rightarrow \lim_{n\to\infty}\left|\frac{(n+1)^{n+1}\,x^{n+1}}{n^n\,x^n}\right| < 1 \Rightarrow |x|\left(\lim_{n\to\infty}\left(1+\frac{1}{n}\right)^n\right)\left(\lim_{n\to\infty}(n+1)\right) < 1 \Rightarrow$

$e\,|x| \lim_{n\to\infty}(n+1) < 1 \Rightarrow$ only $x = 0$ would satisfy this inequality.

a) the radius, 0; this series converges only for $x = 0$

b) this series converges absolutely only for $x = 0$

c) there are no values for which the series is conditionally convergent

27. $\lim_{n\to\infty}\left|\frac{u_{n+1}}{u_n}\right| < 1 \Rightarrow \lim_{n\to\infty}\left|\frac{x^{n+1}}{(n+1)\left(\ln(n+1)\right)^2}\cdot\frac{n(\ln n)^2}{x^n}\right| < 1 \Rightarrow |x|\left(\lim_{n\to\infty}\frac{n}{n+1}\right)\left(\lim_{n\to\infty}\frac{\ln n}{\ln(n+1)}\right)^2 < 1 \Rightarrow$

$|x|\,(1)\left(\lim_{n\to\infty}\frac{1/n}{1/(n+1)}\right)^2 < 1 \Rightarrow |x|\left(\lim_{n\to\infty}\frac{n+1}{n}\right)^2 < 1 \Rightarrow |x|\,(1)^2) < 1 \Rightarrow |x| < 1 \Rightarrow -1 < x < 1$, when $x = -1$

we have $\sum_{n=1}^{\infty}\frac{(-1)^n}{n(\ln n)^2}$ which converges absolutely, where $x = 1$ we have $\sum_{n=1}^{\infty}\frac{1}{n(\ln n)^2}$, which converges.

a) the radius, 1; the interval of convergence, $-1 \le x \le 1$

b) the interval of absolute convergence, $-1 \le x \le 1$

c) there are no values for which the series is conditionally convergent

31. $\lim_{n\to\infty}\left|\frac{u_{n+1}}{u_n}\right| < 1 \Rightarrow \lim_{n\to\infty}\left|\frac{(x+\pi)^{n+1}}{\sqrt{n+1}}\cdot\frac{\sqrt{n}}{(x+\pi)^n}\right| < 1 \Rightarrow |x+\pi|\lim_{n\to\infty}\left|\sqrt{\frac{n}{n+1}}\right| < 1 \Rightarrow$

$|x+\pi|\sqrt{\lim_{n\to\infty}\frac{n}{n+1}} < 1 \Rightarrow |x+\pi|\sqrt{1} < 1 \Rightarrow |x+\pi| < 1 \Rightarrow -1 < x+\pi < 1 \Rightarrow -1-\pi < x < 1-\pi$, when

$x = -1-\pi$ we have $\sum_{n=1}^{\infty}\frac{(-1)^n}{\sqrt{n}} = \sum_{n=1}^{\infty}\frac{(-1)^n}{n^{1/2}}$ which converges by the Alternating Series Theorem but

$\sum_{n=1}^{\infty}\left|\frac{(-1)^n}{n^{1/2}}\right| = \sum_{n=1}^{\infty}\frac{1}{n^{1/2}}$ a divergent p–series, when $x = 1-\pi$ we have $\sum_{n=1}^{\infty}\frac{1^n}{\sqrt{n}} = \sum_{n=1}^{\infty}\frac{1}{n^{1/2}}$ a divergent p–series.

a) the radius, 1; the interval of convergence, $(-1-\pi) \le x < (1-\pi)$

b) the interval of absolute convergence, $-1-\pi < x < 1-\pi$

c) This series is conditionally convergent at $x = -1-\pi$.

35. $\lim_{n\to\infty}\left|\frac{u_{n+1}}{u_n}\right| < 1 \Rightarrow \lim_{n\to\infty}\left|\frac{(\sqrt{x}-2)^{n+1}}{2^{n+1}}\cdot\frac{2^n}{(\sqrt{x}-2)^n}\right| < 1 \Rightarrow \left|\sqrt{x}-2\right| < 2 \Rightarrow -2 < \sqrt{x}-2 < 2 \Rightarrow 0 < \sqrt{x} < 4 \Rightarrow$

$0 < x < 16$, when $x = 0$ we have $\sum_{n=0}^{\infty}(-1)^n$ a divergent series, when $x = 16$ we have $\sum_{n=0}^{\infty}(1)^n$ a divergent series,

therefore the interval of convergence is $0 < x < 16$. The series $\sum_{n=0}^{\infty}\left(\frac{\sqrt{x}-2}{2}\right)^n$ is a convergent geometric series

when $0 < x < 16$ and its sum is $= \frac{1}{1-\frac{\sqrt{x}-2}{2}} = \frac{1}{\frac{2-\sqrt{x}+2}{2}} = \frac{1}{\frac{4-\sqrt{x}}{2}} = \frac{2}{4-\sqrt{x}}$

39. $\lim_{n\to\infty}\left|\frac{(x-3)^{n+1}}{2^{n+1}}\ \frac{2^n}{(x-3)^n}\right| < 1 \Rightarrow |x-3| < 2 \Rightarrow 1 < x < 5$; both series: $\sum_{n=1}^{\infty}(1)$, when $x = 1$ and $\sum_{n=1}^{\infty}(-1)^n$,

when $x = 5$ diverge and therefore the interval of convergence is $1 < x < 5$. The sum of this convergent geometric series is $\frac{1}{1+\frac{x-3}{2}} = \frac{2}{x-1}$. $f(x) = 1 - \frac{1}{2}(x-3) + \frac{1}{4}(x-3)^2 + \ldots + \left(-\frac{1}{2}\right)^n(x-3)^n + \ldots = \frac{2}{x-1} \Rightarrow f'(x) = -\frac{1}{2} + \frac{1}{2}(x-3) + \ldots + \left(-\frac{1}{2}\right)^n n(x-3)^{n-1} + \ldots$ is convergent when $1 < x < 5$, and diverges when $x = 1$ or 5. The sum is $\frac{-2}{(x-1)^2}$ the derivative of $\frac{2}{x-1}$.

43. a) If $f(x) = \sum_{n=0}^{\infty} a_n x^n$, then $f^{(k)}(x) = \sum_{n=k}^{\infty} n(n-1)(n-2)\cdots(n-(k-1))\, a_n x^{n-k}$ and $f^{(k)}(0) = k!\, a_k \Rightarrow$

$a_k = \frac{f^{(k)}(0)}{k!}$ and likewise if $f(x) = \sum_{n=0}^{\infty} b_n x^n$, then $b_k = \frac{f^{(k)}(0)}{k!} \Rightarrow a_k = b_k$ for every nonnegative integer k.

b) If $f(x) = \sum_{n=0}^{\infty} a_n x^n = 0$ for all x, then $f^{(k)}(x) = 0$ for all x and from part a) $a_k = 0$ for every nonnegative integer k.

47. The given series converges for $-\infty < x < \infty$, so its derivative must converge for $-\infty < x < \infty$ by Theorem 12. The derivative is identical with the original series, so $f'(x) = f(x)$. This means that $f(x) = C e^x$ for some constant C. But $f(0) = 1$, so $C e^0 = C \cdot 1$ and $C = 1$. Hence $f(x) = e^x$

9.7 TAYLOR AND MACLAURIN SERIES

1. $f(x) = \ln x$, $f'(x) = \frac{1}{x}$, $f''(x) = -\frac{1}{x^2}$, $f'''(x) = \frac{2}{x^3}$, $f(1) = \ln 1 = 0$, $f'(1) = 1$, $f''(1) = -1$, $f'''(1) = 2 \Rightarrow P_0(x) = 0$,
$P_1(x) = x - 1$, $P_2(x) = (x-1) - \frac{1}{2}(x-1)^2$, $P_3(x) = (x-1) - \frac{1}{2}(x-1)^2 + \frac{1}{3}(x-1)^3$

5. $f(x) = \sin x \Rightarrow f'(x) = \cos x \Rightarrow f''(x) = -\sin x \Rightarrow f'''(x) = -\cos x \Rightarrow f(\pi/4) = \sin \pi/4 = \frac{\sqrt{2}}{2}$, $f'(\pi/4) = \cos \pi/4 = \frac{\sqrt{2}}{2}$,
$f''(\pi/4) = -\sin \pi/4 = -\frac{\sqrt{2}}{2}$, $f'''(\pi/4) = -\cos \pi/4 = -\frac{\sqrt{2}}{2} \Rightarrow P_0 = \frac{\sqrt{2}}{2}$, $P_1(x) = \frac{\sqrt{2}}{2} + \frac{\sqrt{2}}{2}\left(x - \frac{\pi}{4}\right)$,
$P_2(x) = \frac{\sqrt{2}}{2} + \frac{\sqrt{2}}{2}\left(x - \frac{\pi}{4}\right) - \frac{\sqrt{2}}{4}\left(x - \frac{\pi}{4}\right)^2$, $P_3(x) = \frac{\sqrt{2}}{2} + \frac{\sqrt{2}}{2}\left(x - \frac{\pi}{4}\right) - \frac{\sqrt{2}}{4}\left(x - \frac{\pi}{4}\right)^2 - \frac{\sqrt{2}}{12}\left(x - \frac{\pi}{4}\right)^3$

9. $e^x = \sum_{n=0}^{\infty} \frac{x^n}{n!} \Rightarrow e^{-5x} = \sum_{n=0}^{\infty} \frac{(-5x)^n}{n!} = 1 - 5x + \frac{25x^2}{2!} - \frac{125x^3}{3!} + \frac{625x^4}{4!} - \ldots$

13. $7\cos(-x) = 7\cos(x) = 7 \sum_{n=0}^{\infty} \frac{(-1)^n x^{2n}}{(2n)!} = 7 - \frac{7x^2}{2!} + \frac{7x^4}{4!} - \frac{7x^6}{6!} + \ldots$, since cosine is an even function

17. $e^x = \sum_{n=0}^{\infty} \frac{x^n}{n!} \Rightarrow x e^x = x\left(\sum_{n=0}^{\infty} \frac{x^n}{n!}\right) = \sum_{n=0}^{\infty} \frac{x^{n+1}}{n!} = x + x^2 + \frac{x^3}{2!} + \frac{x^4}{3!} + \frac{x^5}{4!} + \ldots$

21. $\cos x = \sum_{n=0}^{\infty} \frac{(-1)^n x^{2n}}{(2n)!} \Rightarrow \frac{x^2}{2} - 1 + \cos x = \frac{x^2}{2} - 1 + \sum_{n=0}^{\infty} \frac{(-1)^n x^{2n}}{(2n)!} = \frac{x^2}{2} - 1 + 1 - \frac{x^2}{2} + \frac{x^4}{4!} - \frac{x^6}{6!} + \frac{x^8}{8!} - \frac{x^{10}}{10!} + \ldots =$
$\frac{x^4}{4!} - \frac{x^6}{6!} + \frac{x^8}{8!} - \frac{x^{10}}{10!} + \ldots$

25. $\cosh x = \frac{e^x + e^{-x}}{2} = \frac{1}{2}\left[\left(1 + x + \frac{x^2}{2!} + \frac{x^3}{3!} + \frac{x^4}{4!} + \ldots\right) + \left(1 - x + \frac{x^2}{2!} - \frac{x^3}{3!} + \frac{x^4}{4!} - \ldots\right)\right] = 1 + \frac{x^2}{2!} + \frac{x^4}{4!} + \frac{x^6}{6!} + \ldots$

29. $f(x) = \ln(\cos x) \Rightarrow f'(x) = -\frac{\sin x}{\cos x} = -\tan x \Rightarrow f''(x) = -\sec^2 x \Rightarrow f'''(x) = -2\sec^2 x \tan x$. $\therefore \ln(\cos x) = -\frac{x^2}{2} + \frac{f'''(c)}{3!}x^3$ where c is between 0 and x.

33. If $e^x = \sum_{n=0}^{\infty} \frac{f^{(n)}(a)}{n!}(x-a)^n$ and $f(x) = e^x$, we have $f^{(n)}(a) = e^a$ for all n = 0, 1, 2, 3,...;

$e^x = e^a\left[\frac{(x-a)^0}{0!} + \frac{(x-a)^1}{1!} + \frac{(x-a)^2}{2!} + \cdots\right] = e^a\left[1 + (x-a) + \frac{(x-a)^2}{2!} + \cdots\right]$, at x = a.

37. $\sin x = x + R_1(x)$, when $|x| < 10^{-3} \Rightarrow |R_1(x)| = \left|\frac{-\cos c}{3!}x^3\right| < \left|\frac{(1)\,x^3}{3!}\right| < \frac{\left(10^{-3}\right)^3}{3!} = 1.67 \times 10^{-10}$

From exercise 51 in section 9.5, $R_1(x)$ has the same sign as $-\frac{x^3}{3!}$. $x < \sin x \Rightarrow 0 < \sin x - x = R_1(x)$, which has the same sign as $-\frac{x^3}{3!} \Rightarrow x < 0 \Rightarrow -10^{-3} < x < 0$.

41. sin x, when x = 0.1; the sum is $\sin(0.1) \approx 0.099833416$

45. $2[\cos x][\sin x] = 2\left[1 - \frac{x^2}{2!} + \frac{x^4}{4!} - \frac{x^6}{6!} + \cdots\right]\left[x - \frac{x^3}{3!} + \frac{x^5}{5!} - \frac{x^7}{7!} + \cdots\right] = 2\left[x - \frac{4x^3}{3!} + \frac{16x^5}{5!} - \frac{64x^7}{7!} + \frac{256x^9}{9!} - \cdots\right] = 2x - \frac{8x^3}{3!} + \frac{32x^5}{5!} - \frac{128\,x^7}{7!} + \frac{512\,x^9}{9!} - \cdots = \sin 2x$

49. $\sin^2 x = \left(\frac{1-\cos 2x}{2}\right) = \frac{1}{2} - \frac{1}{2}\cos 2x = \frac{1}{2} - \frac{1}{2}\left(1 - \frac{(2x)^2}{2!} + \frac{(2x)^4}{4!} - \frac{(2x)^6}{6!} + \cdots\right) = \frac{2\,x^2}{2!} - \frac{2^3\,x^4}{4!} + \frac{2^5\,x^6}{6!} - \cdots$,
$\frac{d}{dx}\left(\sin^2 x\right) = \frac{d}{dx}\left(\frac{2\,x^2}{2!} - \frac{2^3\,x^4}{4!} + \frac{2^5\,x^6}{6!} - \cdots\right) = 2x - \frac{(2x)^3}{3!} + \frac{(2x)^5}{5!} - \frac{(2x)^7}{7!} + \cdots = \sin 2x$

53. $e^x = 1 + x + \frac{x^2}{2!} + \frac{x^3}{3!} + \frac{x^4}{4!} + \cdots \Rightarrow e^{i\theta} = 1 + i\theta + \frac{(i\theta)^2}{2!} + \frac{(i\theta)^3}{3!} + \frac{(i\theta)^4}{4!} + \cdots$ and $e^{-i\theta} = 1 - i\theta + \frac{(-i\theta)^2}{2!} + \frac{(-i\theta)^3}{3!} + \frac{(-i\theta)^4}{4!} + \cdots = 1 - i\theta + \frac{(i\theta)^2}{2!} - \frac{(i\theta)^3}{3!} + \frac{(i\theta)^4}{4!} - \cdots$, $\frac{e^{i\theta} + e^{-i\theta}}{2} =$

$$\frac{\left(1 + i\theta + \frac{(i\theta)^2}{2!} + \frac{(i\theta)^3}{3!} + \frac{(i\theta)^4}{4!} + \cdots\right) + \left(1 - i\theta + \frac{(i\theta)^2}{2!} - \frac{(i\theta)^3}{3!} + \frac{(i\theta)^4}{4!} - \cdots\right)}{2} = 1 - \frac{\theta^2}{2!} + \frac{\theta^4}{4!} - \frac{\theta^6}{6!} + \cdots = \cos\theta;$$

$$\frac{e^{i\theta} - e^{-i\theta}}{2} = \frac{\left(1 + i\theta + \frac{(i\theta)^2}{2!} + \frac{(i\theta)^3}{3!} + \frac{(i\theta)^4}{4!} + \cdots\right) - \left(1 - i\theta + \frac{(i\theta)^2}{2!} - \frac{(i\theta)^3}{3!} + \frac{(i\theta)^4}{4!} - \cdots\right)}{2i} = \theta - \frac{\theta^3}{3!} + \frac{\theta^5}{5!} - \frac{\theta^7}{7!} + \cdots = \sin\theta$$

57. If $f(x) = \sum_{n=0}^{\infty} a_n x^n$, then $f^{(k)}(x) = \sum_{n=k}^{\infty} n(n-)(n-2)\cdots(n)(k-1)\; a_k\, x^{n-k}$ and $f^{(k)}(0) = k!\, a_k \Rightarrow a_k = \frac{f^{(k)}(0)}{k!}$ for k a nonnegative integer. $\therefore$ the coefficients of f(x) are identical with the corresponding coefficients in the Maclaurin series of f(x) and the statements follow.

9.8 CALCULATIONS WITH TAYLOR SERIES

1. $(1+x)^{1/2} = 1^{1/2} + \frac{(1/2)(1)^{-1/2}x}{1!} + \frac{(1/2)(-1/2)(1)^{-3/2}x^2}{2!} + \frac{(1/2)(-1/2)(-3/2)(1)^{-5/2}x^3}{3!} + \ldots$. The first four terms are $1, \frac{x}{2}, -\frac{x^2}{8}, \frac{x^3}{16}$.

5. $(1+x/2)^{-2} = 1^{-2} + \frac{(-2)(1)^{-3}(x/2)}{1!} + \frac{(-2)(-3)(1)^{-4}(x/2)^2}{2!} + \frac{(-2)(-3)(-4)(1)^{-5}(x/2)^3}{3!} + \ldots$. The first four terms are $1, -x, \frac{3x^2}{4}, -\frac{x^3}{2}$.

9. $(1+1/x)^{1/2} = 1^{1/2} + \frac{(1/2)(1)^{-1/2}(1/x)}{1!} + \frac{(1/2)(-1/2)(1)^{-3/2}(1/x)^2}{2!} + \frac{(1/2)(-1/2)(-3/2)(1)^{-5/2}(1/x)^3}{3!} + \ldots$. The first four terms are $1, \frac{1}{2x}, -\frac{1}{8x^2}, \frac{1}{16x^3}$.

13. $(1-2x)^3 = (1+(-2x))^3 = (1)^3 + \frac{(3)(1)^2(-2x)^1}{1!} + \frac{(3)(2)(1)^1(-2x)^2}{2!} + \frac{(3)(2)(1)(1)^0(-2x)^3}{3!} = 1 - 6x + 12x^2 - 8x^3$.

17. $\int_0^{0.1} x^2 e^{-x^2}\,dx = \int_0^{0.1} x^2\left(1 - x^2 + \frac{x^4}{2!} - \frac{x^6}{3!} + \cdots\right)dx = \int_0^{0.1}\left(x^2 - x^4 + \frac{x^6}{2!} - \cdots\right)dx =$
$\left[\frac{x^3}{3} - \frac{x^5}{5} + \cdots\right]_0^{0.1} \approx \left[\frac{x^3}{3}\right]_0^{0.1} \approx 0.00033$ with error $|E| \le \frac{(0.1)^5}{5} \approx 0.000002$

21. $\int_0^{0.1} \frac{1}{\sqrt{1+x^4}}\,dx = \int_0^{0.1}\left(1 - \frac{x^4}{2} + \frac{3x^8}{8} - \cdots\right)dx = \left[x - \frac{x^5}{10} + \cdots\right]_0^{0.1} \approx [x]_0^{0.1} \approx 0.1$ with error

$|E| \le \frac{(0.1)^5}{10} = 0.000001$

25. $(1+x^4)^{1/2} = (1)^{1/2} + \frac{1/2}{1}(1)^{-1/2}(x^4) + \frac{(1/2)(-1/2)}{2!}(1)^{-3/2}(x^4)^2 + \frac{(1/2)(-1/2)(-3/2)}{3!}(1)^{-5/2}(x^4)^3 +$
$\frac{(1/2)(-1/2)(-3/2)(-5/2)}{4!}(1)^{-7/2}(x^4)^4 + \ldots = 1 + \frac{x^4}{2} - \frac{x^8}{8} + \frac{x^{12}}{16} - \frac{5x^{16}}{128} + \ldots;$
$\int_0^{0.1} 1 + \frac{x^4}{2} - \frac{x^8}{8} + \frac{x^{12}}{16} - \frac{5x^{16}}{128} + \ldots dx = \left[x + \frac{x^5}{10} - \frac{x^9}{72} + \frac{x^{13}}{208} - \frac{5x^{17}}{2176} + \ldots\right]_0^{0.1} \approx 0.100001$

29. $\tan^{-1}x = x - \frac{x^3}{3} + \frac{x^5}{5} - \frac{x^7}{7} + \frac{x^9}{9} - \ldots + \frac{(-1)^{n-1}x^{2n-1}}{2n-1} + \ldots$ and the $|\text{error}| = \left|\frac{(-1)^{n-1}x^{2n-1}}{2n-1}\right| = \frac{1}{2n-1}$, when $x = 1$; $\frac{1}{2n-1} < \frac{1}{10^3} \Rightarrow n > \frac{1001}{2} = 500.5 \Rightarrow$ the first term not used is $501^{st} \Rightarrow$ we must use 500 terms

33. a) $(1-x^2)^{-1/2} \approx 1 + \frac{x^2}{2} + \frac{3x^4}{8} + \frac{5x^6}{16} \Rightarrow \sin^{-1}x \approx x + \frac{x^3}{6} + \frac{3x^5}{40} + \frac{5x^7}{112}$;
$\lim_{n\to\infty}\left|\frac{1\cdot3\cdot5\cdots(2n-1)(2n+1)x^{2n+3}}{2\cdot4\cdot6\cdots(2n)(2n+2)(2n+3)} \cdot \frac{2\cdot4\cdot6\cdots(2n)(2n+1)}{1\cdot3\cdot5\cdots(2n-1)x^{2n+1}}\right| < 1 \Rightarrow x^2 \lim_{n\to\infty}\left|\frac{(2n+1)(2n+1)}{(2n+2)(2n+3)}\right| < 1 \Rightarrow$
$|x| < 1 \Rightarrow$ the radius of convergence is 1

b) since $\frac{d}{dx}(\cos^{-1}x) = -(1-x^2)^{-1/2} \Rightarrow \cos^{-1}x = \frac{\pi}{2} - \sin^{-1}x \approx \frac{\pi}{2} - \left(x + \frac{x^3}{6} + \frac{3x^5}{40} + \frac{5x^7}{112}\right) \approx \frac{\pi}{2} - x - \frac{x^3}{6} - \frac{3x^5}{40} - \frac{5x^7}{112}$

37. Assume that the solution has the form $y = a_0 + a_1x + a_2x^2 + \cdots + a_{n-1}x^{n-1} + a_nx^n + \cdots$. $\frac{dy}{dx} = a_1 + 2a_2x + \cdots + na_nx^{n-1} + \cdots$. Now $\frac{dy}{dx} + y = (a_1 + a_0) + (2a_2 + a_1)x + (3a_3 + a_2)x^2 + \cdots + (na_n + a_{n-1})x^{n-1} + \cdots = 0 \Rightarrow$ $a_1 + a_0 = 0$, $2a_2 + a_1 = 0$, $3a_3 + a_2 = 0$ and in general $na_n + a_{n-1} = 0$. Since $y = 1$ when $x = 0$ we have $a_0 = 1$. Therefore $a_1 = -1$, $a_2 = \frac{-a_1}{2 \cdot 1} = \frac{1}{2}$, $a_3 = \frac{-a_2}{3} = -\frac{1}{3 \cdot 2}, \ldots, a_n = \frac{-a_{n-1}}{n} = \frac{(-1)^n}{n!} \Rightarrow y = 1 - x + \frac{1}{2}x^2 - \frac{1}{3 \cdot 2}x^3 + \cdots + \frac{(-1)^n}{n!}x^n + \cdots = \sum_{n=0}^{\infty} \frac{(-1)^n x^n}{n!} = e^{-x}$.

41. Assume that the solution has the form $y = a_0 + a_1x + a_2x^2 + \cdots + a_{n-1}x^{n-1} + a_nx^n + \cdots$. $\frac{dy}{dx} = a_1 + 2a_2x + \cdots + na_nx^{n-1} + \cdots$. Now $\frac{dy}{dx} - 2y = (a_1 - 2a_0) + (2a_2 - 2a_1)x + (3a_3 - 2a_2)x^2 + \cdots + (na_n - 2a_{n-1})x^{n-1} + \cdots = 0$. Since $y = 1$ when $x = 0$ we have $a_0 = 1$. Therefore $a_1 = 2a_0 = 2(1) = 2$, $a_2 = \frac{2}{2}a_1 = \frac{2}{2}(2) = \frac{2^2}{2}$, $a_3 = \frac{2}{3}a_2 = \frac{2}{3}\left(\frac{2^2}{2}\right) = \frac{2^3}{3 \cdot 2}, \ldots, a_n = \frac{2}{n}a_{n-1} = \frac{2}{n}\left(2^{n-1}\right) = \frac{2^n}{n!} \Rightarrow y = 1 + 2x + \frac{2^2}{2}x^2 + \frac{2^3}{3 \cdot 2}x^3 + \cdots + \frac{2^n}{n!}x^n + \cdots = 1 + (2x) + \frac{(2x)^2}{2!} + \frac{(2x)^3}{3!} + \cdots + \frac{(2x)^n}{n!} + \cdots = \sum_{n=0}^{\infty} \frac{(2x)^n}{n!} = e^{2x}$.

45. Assume that the solution has the form $y = a_0 + a_1x + a_2x^2 + \cdots + a_{n-1}x^{n-1} + a_nx^n + \cdots$. $\frac{dy}{dx} = a_1 + 2a_2x + \cdots + na_nx^{n-1} + \cdots$. Now $\frac{dy}{dx} + y = (a_1 + a_0) + (2a_2 + a_1)x + (3a_3 + a_2)x^2 + \cdots + (na_n + a_{n-1})x^{n-1} + \cdots = x \Rightarrow$ $a_1 + a_0 = 0$, $2a_2 + a_1 = 1$, $3a_3 + a_2 = 0$ and in general $na_n + a_{n-1} = 0$. Since $y = 0$ when $x = 0$ we have $a_0 = 0$. Therefore $a_1 = 0$, $a_2 = \frac{1 - a_1}{2} = \frac{1}{2}$, $a_3 = \frac{-a_2}{3} = -\frac{1}{3 \cdot 2}, \ldots, a_n = \frac{-a_{n-1}}{n} = \frac{(-1)^n}{n!} \Rightarrow y = 0 - 0x + \frac{1}{2}x^2 - \frac{1}{3 \cdot 2}x^3 + \cdots + \frac{(-1)^n}{n!}x^n + \cdots = \left[1 - 1x + \frac{1}{2}x^2 - \frac{1}{3 \cdot 2}x^3 + \cdots + \frac{(-1)^n}{n!}x^n + \cdots\right] - 1 + x = \left[\sum_{n=0}^{\infty} \frac{(-1)^n x^n}{n!}\right] - 1 + x = e^{-x} - 1 + x = e^{-x} + x - 1$.

9.P PRACTICE EXERCISES

1. converges to 1, since $\lim_{n \to \infty} a_n = \lim_{n \to \infty} \left(1 + \frac{(-1)^n}{n}\right) = 1$

5. diverges, since $\left\{\sin \frac{n\pi}{2}\right\} = 0, 1, 0, -1, 0, 1, \ldots$

9. converges to 1, since $\lim_{n \to \infty} a_n = \lim_{n \to \infty} \left(\frac{n + \ln n}{n}\right) = \lim_{n \to \infty} \frac{1 + 1/n}{1} = 1$

13. converges to 3, since $\lim_{n\to\infty} a_n = \lim_{n\to\infty} \left(\frac{3^n}{n}\right)^{1/n} = \frac{3}{\lim_{n\to\infty} n^{1/n}} = \frac{3}{1} = 3$ by table 9.1

17. Rewrite $\frac{1}{(2n-3)(2n-1)}$ as $\frac{A}{2n-3} + \frac{B}{2n-1} \Rightarrow \frac{A(2n-1)+B(2n-3)}{(2n-3)(2n-1)} = \frac{1}{(2n-3)(2n-1)} \Rightarrow A(2n-1) + B(2n-3) =$ $1 \Rightarrow (2A + 2B)n + (-A - 3B) = 1 \Rightarrow \begin{cases} A+B=0 \\ -A-3B=1 \end{cases} \Rightarrow -2B = 1 \Rightarrow B = -1/2$ and $A = 1/2$.

$$s_n = \sum_{k=3}^{n} \frac{1}{(2k-3)(2k-1)} = \sum_{k=3}^{n} \left(\frac{1/2}{2k-3} + \frac{-1/2}{2k-1}\right) = \frac{1}{2} \sum_{k=3}^{n} \left(\frac{1}{2k-3} - \frac{1}{2k-1}\right) =$$

$\frac{1}{2}\left[\frac{1}{3} - \frac{1}{5} + \frac{1}{5} - \frac{1}{7} + \frac{1}{7} - \frac{1}{9} + \cdots + \frac{1}{2n-5} - \frac{1}{2n-3} + \frac{1}{2n-3} - \frac{1}{2n-1}\right] = \frac{1}{2}\left[\frac{1}{3} - \frac{1}{2n-1}\right]$. The sum is $\lim_{n\to\infty} s_n =$ $\lim_{n\to\infty} \frac{1}{2}\left[\frac{1}{3} - \frac{1}{2n-1}\right] = \frac{1}{6}$.

21. $\sum_{n=0}^{\infty} e^{-n} = \sum_{n=0}^{\infty} \frac{1}{e^n}$ a convergent geometric series with $r = \frac{1}{e}$ and $a = 1$. The sum is $\frac{1}{1-\frac{1}{e}} = \frac{1}{\frac{e-1}{e}} = \frac{e}{e-1}$.

25. Since, $f(x) = \frac{1}{x^{1/2}} \Rightarrow f'(x) = -\frac{1}{2x^{3/2}} < 0 \Rightarrow f(x)$ is decreasing $\Rightarrow a_{n+1} < a_n$ and $\lim_{n\to\infty} a_n = \lim_{n\to\infty} \frac{(-1)^n}{\sqrt{n}} = 0$,

the $\sum_{n=1}^{\infty} \frac{(-1)^n}{\sqrt{n}}$ converges, by the Alternating Series Theorem. The series $\sum_{n=1}^{\infty} \frac{1}{\sqrt{n}}$ diverges $\Rightarrow$ the given series converges conditionally.

29. converges absolutely by the Direct Comparison Test for $\frac{\ln n}{n^3} < \frac{n}{n^3} = \frac{1}{n^2}$ the nth term of a convergent p–series

33. converges absolutely, by the Ratio Test, since $\lim_{n\to\infty} \left|\frac{n+2}{(n+1)!} \cdot \frac{n!}{n+1}\right| = \lim_{n\to\infty} \frac{n+2}{(n+1)^2} = 0 < 1$

37. converges absolutely, since $\lim_{n\to\infty} \frac{\frac{1}{n^{3/2}}}{\frac{1}{\sqrt{n(n+1)(n+2)}}} = \sqrt{\lim_{n\to\infty} \frac{n(n+1)(n+2)}{n^3}} = 1$, by the Limit Comparison Test

41. $\lim_{n\to\infty} \left|\frac{u_{n+1}}{u_n}\right| < 1 \Rightarrow \lim_{n\to\infty} \left|\frac{(3x-1)^{n+1}}{(n+1)^2} \cdot \frac{n^2}{(3x-1)^n}\right| < 1 \Rightarrow |3x-1| \lim_{n\to\infty} \frac{n^2}{(n+1)^2} < 1 \Rightarrow |3x-1|\,(1) < 1 \Rightarrow$

$|3x-1| < 1 \Rightarrow -1 < 3x - 1 < 1 \Rightarrow 0 < 3x < 2 \Rightarrow 0 < x < 2/3$, at $x = 0$ we have $\sum_{n=1}^{\infty} \frac{(-1)^{n-1}(-1)^n}{n^2} =$

$\sum_{n=1}^{\infty} \frac{(-1)^{2n-1}}{n^2} = -\sum_{n=1}^{\infty} \frac{1}{n^2}$, the product of a nonzero constant and a convergent p–series is a convergent series which Is also absolutely convergent; at x = 2/3 we have $\sum_{n=1}^{\infty} \frac{(-1)^{n-1}(1)^n}{n^2} = \sum_{n=1}^{\infty} \frac{(-1)^{n-1}}{n^2}$ an absolutely convergent series.

a) the radius, 1/3; the interval of convergence, $0 \le x \le 2/3$

b) the interval of absolute convergence, $0 \le x \le 2/3$

c) there are no values for which the series is conditionally converent

45. $\lim_{n \to \infty} \left|\frac{u_{n+1}}{u_n}\right| < 1 \Rightarrow \lim_{n \to \infty} \left|\frac{(n+2)\,x^{2n+1}}{3^{n+1}} \cdot \frac{3^n}{(n+1)\,x^{2n-1}}\right| < 1 \Rightarrow \frac{x^2}{3} \lim_{n \to \infty} \left|\frac{n+2}{n+1}\right| < 1 \Rightarrow -\sqrt{3} < x < \sqrt{3}$; when $x = \pm\sqrt{3}$ both series $\sum_{n=1}^{\infty} -\frac{n+1}{\sqrt{3}}$ and $\sum_{n=1}^{\infty} \frac{n+1}{\sqrt{3}}$ diverge.

a) the radius, $\sqrt{3}$, the interval of convergence, $-\sqrt{3} < x < \sqrt{3}$

b) the interval of absolute convergence, $-\sqrt{3} < x < \sqrt{3}$

c) there are no values for which the series is conditionally convergent

49. The given series is in the form $1 - x + x^2 - x^3 + \ldots + (-x)^n + \ldots = \frac{1}{1+x}$, where $x = \frac{1}{4}$. The sum is $\frac{1}{1 + 1/4} = \frac{4}{5}$.

53. The given series is in the form $1 + x + \frac{x^2}{2!} + \frac{x^3}{3!} + \ldots + \frac{x^n}{n!} + \ldots = e^x$, where x = ln 2. The sum is $e^{\ln(2)} = 2$.

57. Consider $\frac{1}{1-2x}$ as the sum of a convergent geometric series with a = 1 and r = 2x $\Rightarrow \frac{1}{1-2x} = 1 + (2x) + (2x)^2 + (2x)^3 + \ldots = \sum_{n=0}^{\infty} (2x)^n$ where $|2x| < 1 \Rightarrow |x| < \frac{1}{2}$.

61. $\cos x = \sum_{n=0}^{\infty} \frac{(-1)^n x^{2n}}{(2n)!} \Rightarrow \cos\left(x^{5/2}\right) = \sum_{n=0}^{\infty} \frac{(-1)^n \left(x^{5/2}\right)^{2n}}{(2n)!} = \sum_{n=0}^{\infty} \frac{(-1)^n x^{5n}}{(2n)!}$

65. $\int_0^{1/2} \exp\left(-x^3\right) dx = \int_0^{1/2} 1 - x^3 + \frac{x^6}{2!} - \frac{x^9}{3!} + \frac{x^{12}}{4!} + \ldots dx = \left[x - \frac{x^4}{4} + \frac{x^7}{7 \cdot 2!} - \frac{x^{10}}{10 \cdot 3!} + \frac{x^{13}}{13 \cdot 4!} - \ldots\right]_0^{1/2} \approx \frac{1}{2} - \frac{1}{2^4 \cdot 4} + \frac{1}{2^7 \cdot 7 \cdot 2!} - \frac{1}{2^{10} \cdot 10 \cdot 3!} \approx 0.484917151$; note $\exp[f(x)] = e^{f(x)}$

69. Diverges because the nth term, $\left(1 - (1/n)\right)^n$ approaches 1/e as $n \to \infty$. To see why, let $y = \left(1 - (1/n)\right)^n$. Then $\ln y = n \ln\left[\frac{n-1}{n}\right] = \frac{\ln(n-1) - \ln n}{1/n}$ and L'Hôpital's rule gives $\lim_{n \to \infty} \ln y = \lim_{n \to \infty} \frac{\left(1/(n-1)\right) - (1/n)}{-\left(1/n^2\right)} = \lim_{n \to \infty} -\frac{n^2}{n^2 - n} = -1$. Hence $\lim_{n \to \infty} y = e^{-1} = 1/e$.

73. Assume that the solution has the form $y = a_0 + a_1x + a_2x^2 + \cdots + a_{n-1}x^{n-1} + a_nx^n + \cdots$. $\frac{dy}{dx} = a_1 + 2a_2x + \cdots + na_nx^{n-1} + \cdots$. Now $\frac{dy}{dx} + y = (a_1 + a_0) + (2a_2 + a_1)x + (3a_3 + a_2)x^2 + \cdots + (na_n + a_{n-1})x^{n-1} + \cdots = 0 \Rightarrow a_1 + a_0 = 0,\ 2a_2 + a_1 = 0,\ 3a_3 + a_2 = 0$ and in general $na_n + a_{n-1} = 0$. Since $y = -1$ when $x = 0$ we have $a_0 = -1$. Therefore $a_1 = 1$, $a_2 = \frac{-a_1}{2 \cdot 1} = -\frac{1}{2}$, $a_3 = \frac{-a_2}{3} = \frac{1}{3 \cdot 2}$, $a_4 = \frac{-a_3}{4} = -\frac{1}{4 \cdot 3 \cdot 2}$, $\ldots$, $a_n = \frac{-a_{n-1}}{n} = \frac{-1}{n}\frac{(-1)^n}{(n-1)!} = \frac{(-1)^{n+1}}{n!} \Rightarrow y = -1 + x - \frac{1}{2}x^2 + \frac{1}{3 \cdot 2}x^3 - \cdots + \frac{(-1)^{n+1}}{n!}x^n + \cdots =$

$$-1\left[1 - x + \frac{1}{2}x^2 - \frac{1}{3 \cdot 2}x^3 - \cdots + \frac{(-1)^n}{n!}x^n + \cdots\right] = -1\left[\sum_{n=0}^{\infty} \frac{(-1)^n x^n}{n!}\right] = -e^{-x}.$$

CHAPTER 10

CONIC SECTIONS, PARAMETRIZED CURVES, AND POLAR COORDINATES

10.1 CONIC SECTIONS AND QUADRATIC EQUATIONS

1. $x = \frac{y^2}{8} \Rightarrow 4p = 8 \Rightarrow p = 2$. $\therefore$ Focus is (2,0), directrix is $x = -2$.

5. $\frac{x^2}{4} - \frac{y^2}{9} = 1 \Rightarrow c = \sqrt{4+9} = \sqrt{13} \Rightarrow$ Foci are $\left(\pm\sqrt{13}, 0\right)$. $e = \frac{c}{a} = \frac{\sqrt{13}}{2} \Rightarrow \frac{a}{e} = \frac{2}{\frac{\sqrt{13}}{2}} = \frac{4}{\sqrt{13}} \Rightarrow$ Directrices are $x = \pm\frac{4}{\sqrt{13}}$.

   Asymptotes are $y = \pm\frac{3}{2}x$.

9.

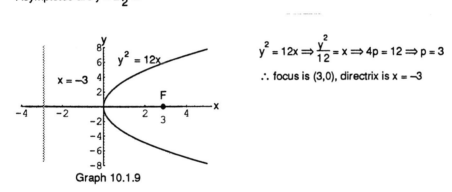

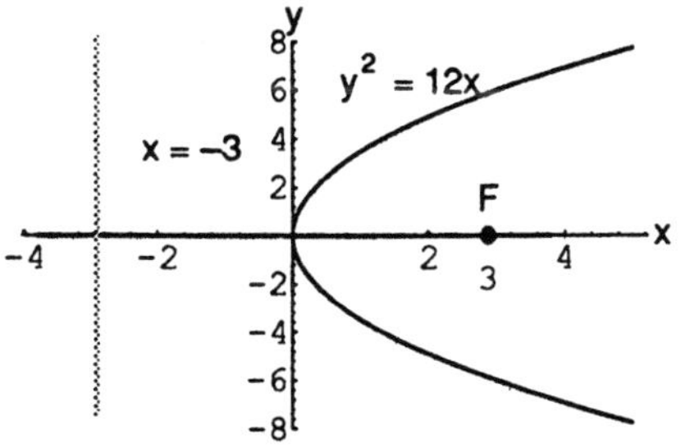

Graph 10.1.9

$y^2 = 12x \Rightarrow \frac{y^2}{12} = x \Rightarrow 4p = 12 \Rightarrow p = 3$

$\therefore$ focus is (3,0), directrix is $x = -3$

13.

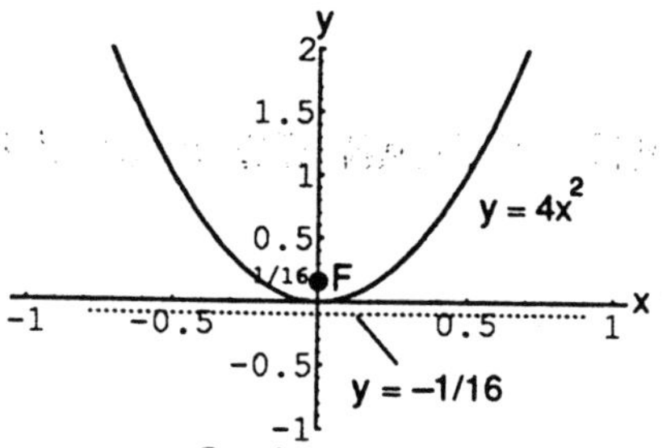

Graph 10.1.13

$y = 4x^2 \Rightarrow y = \frac{x^2}{1/4} \Rightarrow 4p = \frac{1}{4} \Rightarrow p = \frac{1}{16}$.

$\therefore$ focus is $\left(0, \frac{1}{16}\right)$, directrix is $y = -\frac{1}{16}$

17.

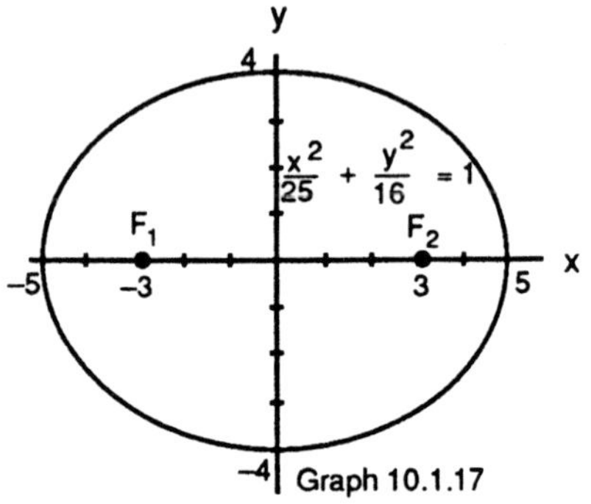

Graph 10.1.17

$16x^2 + 25y^2 = 400 \Rightarrow \frac{x^2}{25} + \frac{y^2}{16} = 1$

$\Rightarrow c = \sqrt{a^2 - b^2} = \sqrt{25 - 16} = 3.$

$e = \frac{c}{a} = \frac{3}{5}$

21.

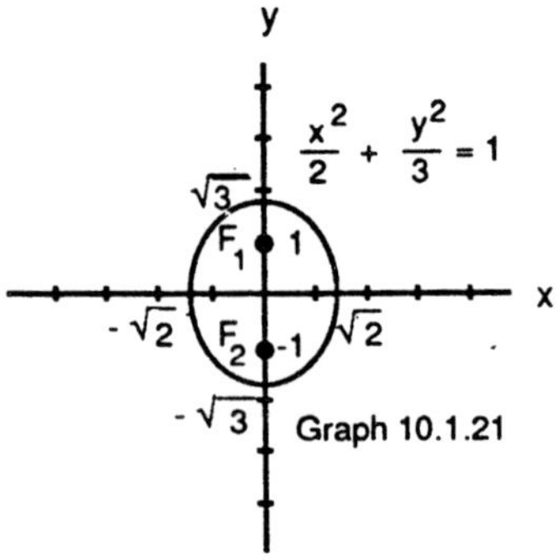

Graph 10.1.21

$3x^2 + 2y^2 = 6 \Rightarrow \frac{x^2}{2} + \frac{y^2}{3} = 1$

$\Rightarrow c = \sqrt{a^2 - b^2} = \sqrt{3 - 2} = 1$

$e = \frac{c}{a} = \frac{1}{\sqrt{3}}$

25.

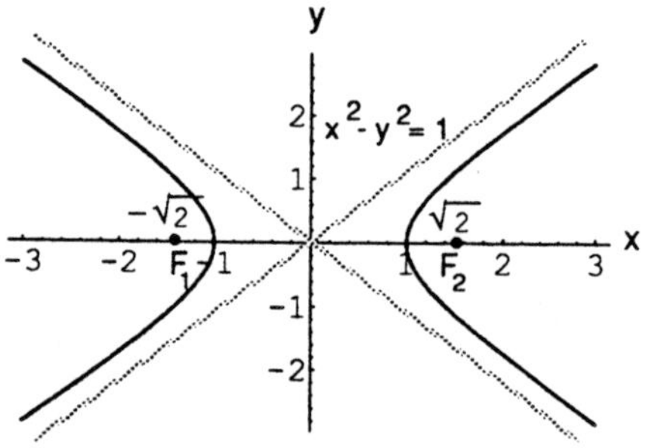

Graph 10.1.25

$x^2 - y^2 = 1 \Rightarrow c = \sqrt{a^2 + b^2} = \sqrt{1 + 1} = \sqrt{2}$

$e = \frac{c}{a} = \frac{\sqrt{2}}{1} = \sqrt{2}$

Asymptotes are $y = \pm x$

29.

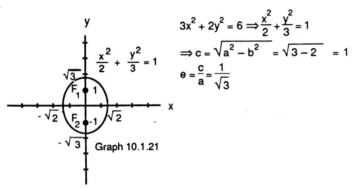

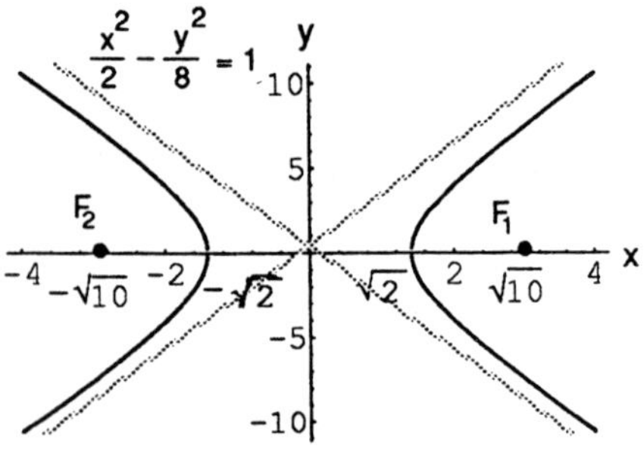

Graph 10.1.29

$8x^2 - 2y^2 = 16 \Rightarrow \frac{x^2}{2} - \frac{y^2}{8} = 1$

$\Rightarrow c = \sqrt{a^2 + b^2} = \sqrt{2 + 8} = \sqrt{10}$

$e = \frac{c}{a} = \frac{\sqrt{10}}{\sqrt{2}} = \sqrt{5}$

Asymptotes are $y = \pm 2x$.

33. Volume of the Parabolic Solid: $V_1 = \int_0^{b/2} 2\pi x\left(h - \frac{4h}{b^2}x^2\right)dx = 2\pi h \int_0^{b/2}\left(x - \frac{4x^3}{b^2}\right)dx = 2\pi h\left[\frac{x^2}{2} - \frac{x^4}{b^2}\right]_0^{b/2} = \frac{\pi h b^2}{8}$

Volume of the Cone: $V_2 = \frac{1}{3}\pi\left(\frac{b}{2}\right)^2 h = \frac{1}{3}\pi\left(\frac{b^2}{4}\right)h = \frac{\pi h b^2}{12} \quad \therefore V_1 = \frac{3}{2}V_2$

37. $9x^2 - 4y^2 = 36,\ x = 4 \Rightarrow y^2 = \frac{9x^2 - 36}{4} \Rightarrow y = \frac{3}{2}\sqrt{x^2 - 4}$

$$V = \int_2^4 \pi\left(\frac{3}{2}\sqrt{x^2 - 4}\right)^2 dx = \frac{9\pi}{4}\int_2^4 \left(x^2 - 4\right) dx = \frac{9\pi}{4}\left[\frac{x^3}{3} - 4x\right]_2^4 = 24\pi$$

41. $\frac{dr_A}{dt} = \frac{dr_B}{dt} \Rightarrow \int \frac{dr_A}{dt} = \int \frac{dr_B}{dt} \Rightarrow r_A + C_1 = r_B + C_2 \Rightarrow r_A - r_B = C$, a constant $\Rightarrow$ The points, P(t), lie on a hyperbola with foci A and B.

45. PF will always equal PB because the string has constant length AB.

49. a) $\frac{x^2}{16} + \frac{y^2}{9} = 1 \Rightarrow$ center is (0,0), vertices are (–4,0), (4,0). b)
$c = \sqrt{a^2 - b^2} = \sqrt{7} \Rightarrow$ foci are $\left(\sqrt{7},\ 0\right), \left(-\sqrt{7},\ 0\right)$
$\therefore$ new center is (4,3), new vertices are (0,3), (8,3), and new foci are $\left(-\sqrt{7} + 4,\ 3\right), \left(\sqrt{7} + 4,\ 3\right)$.

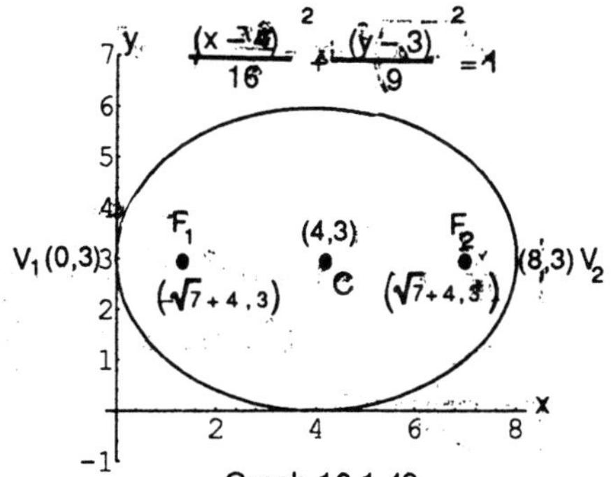

Graph 10.1.49

53. $y^2 = 4x \Rightarrow 4p = 4 \Rightarrow p = 1 \Rightarrow$ focus is (1,0), directrix is $x = -1$. Vertex is (0,0). $\therefore$ new vertex is (–2,–3), new focus is (–1,–3), and new directrix is $x = -3$. The new equation is $(y + 3)^2 = 4(x + 2)$.

57. $\frac{x^2}{6} + \frac{y^2}{9} = 1 \Rightarrow$ center is (0,0), vertices are (0,3), (0,–3). $c = \sqrt{a^2 - b^2} = \sqrt{9 - 6} = \sqrt{3} \Rightarrow$ foci are $\left(0, \sqrt{3}\right), \left(0, -\sqrt{3}\right)$.
$\therefore$ new center is (–2,–1), new vertices are (–2,2), (–2,–4), new foci are $\left(-2, -1 + \sqrt{3}\right), \left(-2, -1 - \sqrt{3}\right)$. The new equation is $\frac{(x + 2)^2}{6} + \frac{(y + 1)^2}{9} = 1$.

61. $\frac{x^2}{4} - \frac{y^2}{5} = 1 \Rightarrow$ center is (0,0), vertices are (2,0), (–2,0). $c = \sqrt{a^2 + b^2} = \sqrt{4 + 5} = 3 \Rightarrow$ foci are (3,0), (–3,0).
Asymptotes are $\pm\frac{x}{2} = \frac{y}{\sqrt{5}} \Rightarrow y = \pm\frac{x\sqrt{5}}{2}$. $\therefore$ new center is (2,2), new vertices are (4,2), (0,2), new foci are (5,2), (–1,2).
The new asymptotes are $y - 2 = \pm\frac{(x - 2)\sqrt{5}}{2}$. The new equation is $\frac{(x - 2)^2}{4} - \frac{(y - 2)^2}{5} = 1$.

65. $x^2 + 5y^2 + 4x = 1 \Rightarrow x^2 + 4x + 4 + 5y^2 = 5 \Rightarrow (x+2)^2 + 5y^2 = 5 \Rightarrow \frac{(x+2)^2}{5} + y^2 = 1$, which is the equation of an ellipse. The center is $(-2,0)$, the vertices are $\left(-2+\sqrt{5}, 0\right), \left(-2-\sqrt{5}, 0\right)$. $c = \sqrt{a^2 - b^2} = \sqrt{5-1} = 2 \Rightarrow$ the foci are $(-4,0)$, $(0,0)$.

69. $x^2 - y^2 - 2x + 4y = 4 \Rightarrow x^2 - 2x + 1 - (y^2 - 4y + 4) = 1 \Rightarrow (x-1)^2 - (y-2)^2 = 1$, which is the equation of a hyperbola. The center is $(1,2)$, the vertices are $(2,2)$, $(0,2)$. $c = \sqrt{a^2 + b^2} = \sqrt{1+1} = \sqrt{2} \Rightarrow$ the foci are $\left(1+\sqrt{2}, 2\right)$, $\left(1-\sqrt{2}, 2\right)$. The asymptotes are $y - 2 = \pm(x-1)$.

10.2 QUADRATIC EQUATIONS IN *X* AND *Y*

1. $x^2 - 3xy + y^2 - x = 0 \Rightarrow B^2 - 4AC = (-3)^2 - 4(1)(1) = 5 > 0 \Rightarrow$ Hyperbola

5. $x^2 + 2xy + y^2 + 2x - y + 2 = 0 \Rightarrow B^2 - 4AC = (2)^2 - 4(1)(1) = 0 \Rightarrow$ Parabola

9. $xy + y^2 - 3x = 5 \Rightarrow B^2 - 4AC = 1^2 - 4(0)(1) = 1 > 0 \Rightarrow$ Hyperbola

13. $x^2 - 3xy + 3y^2 + 6y = 7 \Rightarrow B^2 - 4AC = (-3)^2 - 4(1)(3) = -3 < 0 \Rightarrow$ Ellipse

17. $\cot 2\alpha = \frac{A-C}{B} = \frac{0}{1} = 0 \Rightarrow 2\alpha = \frac{\pi}{2} \Rightarrow \alpha = \frac{\pi}{4}$. $\therefore x = x'\cos\alpha - y'\sin\alpha,\ y = x'\sin\alpha + y'\cos\alpha \Rightarrow$ $x = x'\frac{\sqrt{2}}{2} - y'\frac{\sqrt{2}}{2},\ y = x'\frac{\sqrt{2}}{2} + y'\frac{\sqrt{2}}{2} \Rightarrow \left(\frac{\sqrt{2}}{2}x' - \frac{\sqrt{2}}{2}y'\right)\left(\frac{\sqrt{2}}{2}x' + \frac{\sqrt{2}}{2}y'\right) = 2 \Rightarrow \frac{1}{2}x'^2 - \frac{1}{2}y'^2 = 2 \Rightarrow x'^2 - y'^2 = 4$ $\Rightarrow$ Hyperbola

21. $\cot 2\alpha = \frac{A-C}{B} = \frac{1-1}{-2} = 0 \Rightarrow 2\alpha = \frac{\pi}{2} \Rightarrow \alpha = \frac{\pi}{4}$. $\therefore x = x'\cos\alpha - y'\sin\alpha,\ y = x'\sin\alpha + y'\cos\alpha \Rightarrow$ $x = x'\frac{\sqrt{2}}{2} - y'\frac{\sqrt{2}}{2},\ y = x'\frac{\sqrt{2}}{2} + y'\frac{\sqrt{2}}{2} \Rightarrow \left(\frac{\sqrt{2}}{2}x' - \frac{\sqrt{2}}{2}y'\right)^2 - 2\left(\frac{\sqrt{2}}{2}x' - \frac{\sqrt{2}}{2}y'\right)\left(\frac{\sqrt{2}}{2}x' + \frac{\sqrt{2}}{2}y'\right) + \left(\frac{\sqrt{2}}{2}x' + \frac{\sqrt{2}}{2}y'\right)^2 = 2$ $\Rightarrow y'^2 = 1$, Parallel Horizontal Lines.

25. $\cot 2\alpha = \frac{A-C}{B} = \frac{3-3}{2} = 0 \Rightarrow 2\alpha = \frac{\pi}{2} \Rightarrow \alpha = \frac{\pi}{4}$. $\therefore x = x'\cos\alpha - y'\sin\alpha,\ y = x'\sin\alpha + y'\cos\alpha \Rightarrow$ $x = x'\frac{\sqrt{2}}{2} - y'\frac{\sqrt{2}}{2},\ y = x'\frac{\sqrt{2}}{2} + y'\frac{\sqrt{2}}{2} \Rightarrow 3\left(\frac{\sqrt{2}}{2}x' - \frac{\sqrt{2}}{2}y'\right)^2 + 2\left(\frac{\sqrt{2}}{2}x' - \frac{\sqrt{2}}{2}y'\right)\left(\frac{\sqrt{2}}{2}x' + \frac{\sqrt{2}}{2}y'\right) + 3\left(\frac{\sqrt{2}}{2}x' + \frac{\sqrt{2}}{2}y'\right)^2$ $= 19 \Rightarrow 4x'^2 + 2y'^2 = 19$, Ellipse.

29. $\tan 2\alpha = \frac{-4}{1-4} = \frac{4}{3} \Rightarrow 2\alpha \approx 53.13° \Rightarrow \alpha \approx 26.56° \Rightarrow \sin\alpha \approx 0.45,\ \cos\alpha \approx 0.89$. Then $A' \approx 0.00$, $B' \approx 0.00$, $C' \approx 5.00$, $D' \approx 0$, $E' \approx 0$, and $F' = -5 \Rightarrow 5.00y'^2 - 5 = 0$ or $y' = \pm 1.00$, parallel lines.

33. a) $\frac{x'^2}{b^2} + \frac{y'^2}{a^2} = 1$

b) $\frac{y'^2}{a^2} - \frac{x'^2}{b^2} = 1$

c) $x'^2 + y'^2 = a^2$

d) $y = mx \Rightarrow y - mx = 0 \Rightarrow D = -m,\ E = 1.\ \alpha = 90°$

$\Rightarrow D' = 1, E' = m \Rightarrow my' + x' = 0 \Rightarrow y' = -\frac{1}{m}x'$

e) $y = mx + b \Rightarrow y - mx - b = 0 \Rightarrow D = -m,\ E = 1 \Rightarrow$
$D' = 1,\ E' = m$ (See part d above.) Also $F' = -b$
$\Rightarrow my' + x' - b = 0 \Rightarrow y' = -\frac{1}{m}x' + \frac{b}{m}$.

37. Yes, the graph is a hyperbola. With $AC < 0$ we have $-4AC > 0$ and $B^2 - 4AC > 0$

10.3 PARAMETRIZATIONS OF CURVES

1.

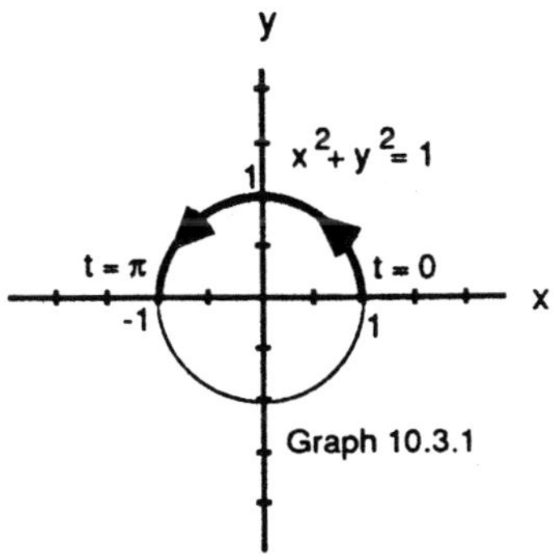

Graph 10.3.1

$x = \cos t,\ y = \sin t,\ 0 \le t \le \pi \Rightarrow$
$\cos^2 t + \sin^2 t = 1 \Rightarrow x^2 + y^2 = 1$

5.

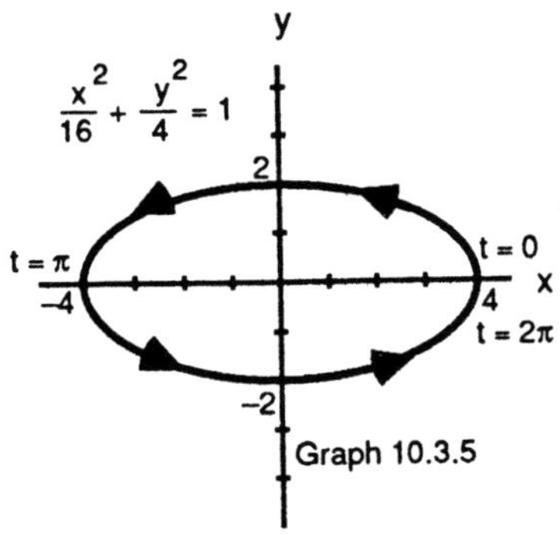

Graph 10.3.5

$x = 4\cos t,\ y = 2\sin t,\ 0 \le t \le 2\pi \Rightarrow$
$\frac{16\cos^2 t}{16} + \frac{4\sin^2 t}{4} = 1 \Rightarrow \frac{x^2}{16} + \frac{y^2}{4} = 1$

9.

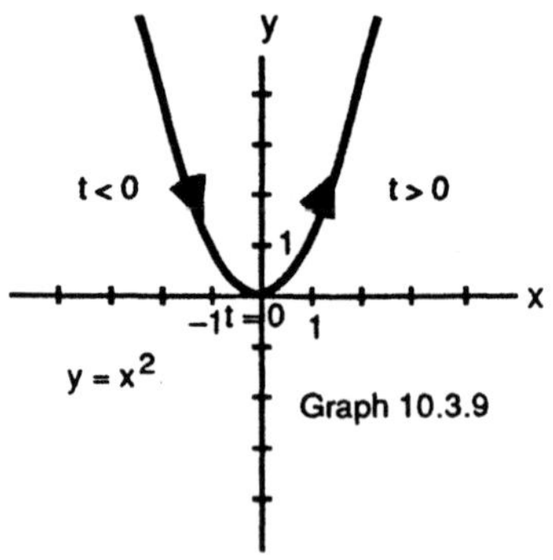

$x = 3t,\ y = 9t^2,\ -\infty < t < \infty \Rightarrow$
$y = x^2$

13.

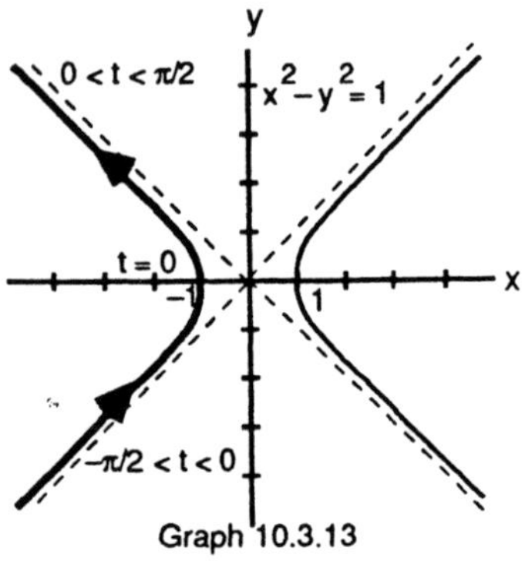

$x = -\sec t,\ y = \tan t,\ -\frac{\pi}{2} < t < \frac{\pi}{2} \Rightarrow$
$\sec^2 t - \tan^2 t = 1 \Rightarrow x^2 - y^2 = 1$

17.

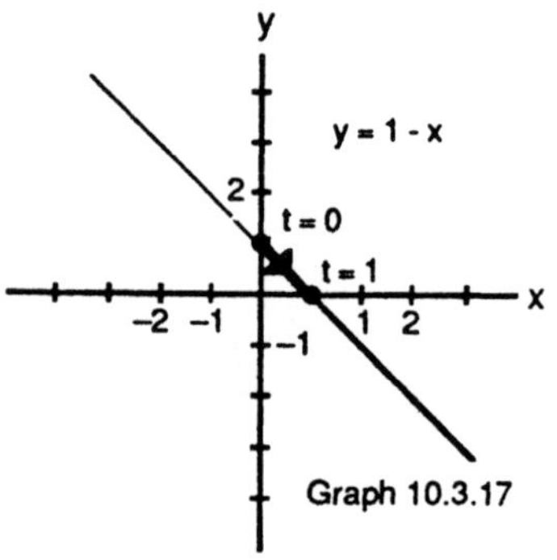

$x = t,\ y = 1 - t,\ 0 \le t \le 1 \Rightarrow$
$y = 1 - x$

21.

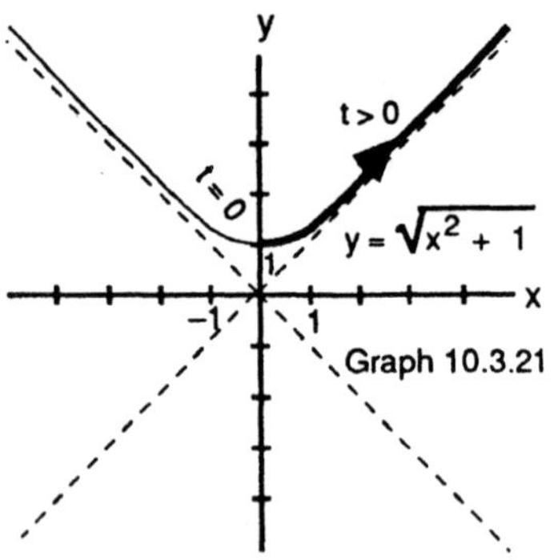

$x = t^2,\ y = \sqrt{t^4 + 1}\ ,\ t \ge 0 \Rightarrow$
$y = \sqrt{x^2 + 1}\ \ ,\ x \ge 0$

25. a) $x = a\cos t,\ y = -a\sin t,\ 0 \le t \le 2\pi$

b) $x = a\cos t,\ y = a\sin t,\ 0 \le t \le 2\pi$

c) $x = a\cos t,\ y = -a\sin t,\ 0 \le t \le 4\pi$

d) $x = a\cos t,\ y = a\sin t,\ 0 \le t \le 4\pi$

29. $d = \sqrt{(x-2)^2 + \left(y - \frac{1}{2}\right)^2} \Rightarrow d^2 = (x-2)^2 + \left(y - \frac{1}{2}\right)^2 = (t-2)^2 + \left(t^2 - \frac{1}{2}\right)^2 \Rightarrow d^2 = t^4 - 4t + \frac{17}{4}$

$\frac{d(d^2)}{dt} = 4t^3 - 4.$ Let $4t^3 - 4 = 0 \Rightarrow t = 1$ The second derivative is always positive for $t \ne 0 \Rightarrow$

$t = 1$ gives a local minimum which is an absolute minimum since it is the only extremum.

$\therefore$ the closest point on the parabola is (1,1).

33. a)

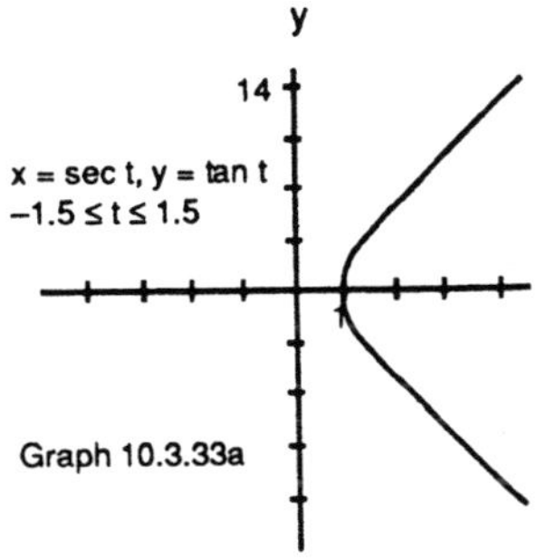

Graph 10.3.33a

b)

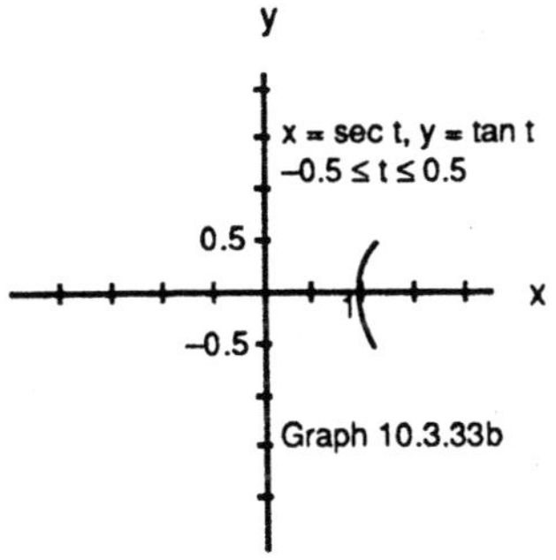

Graph 10.3.33b

c)

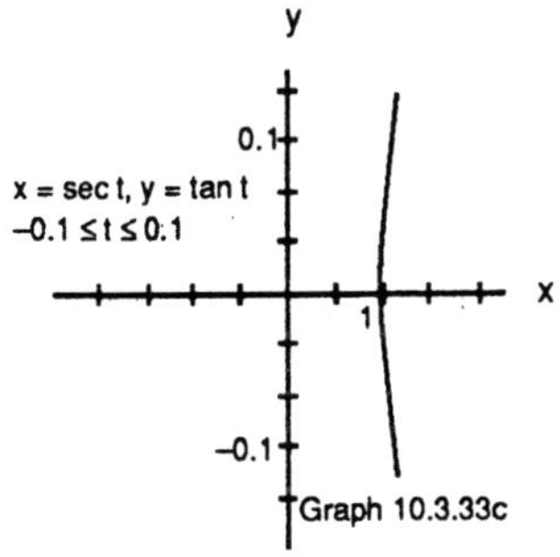

Graph 10.3.33c

37. a)

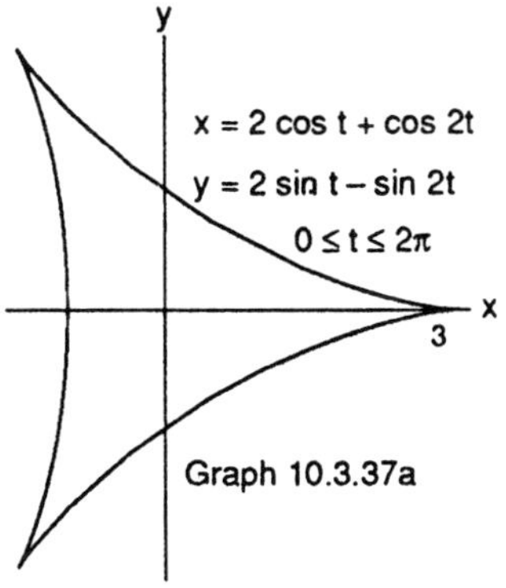

Graph 10.3.37a

b)

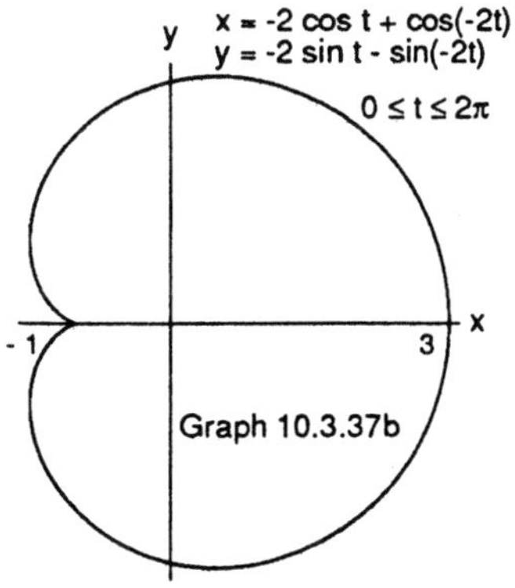

Graph 10.3.37b

10.4 CALCULUS WITH PARAMETRIZED CURVES

1. $t = \frac{\pi}{4} \Rightarrow x = 2\cos\frac{\pi}{4} = \sqrt{2},\ y = 2\sin\frac{\pi}{4} = \sqrt{2}.\ \frac{dx}{dt} = -2\sin t,\ \frac{dy}{dt} = 2\cos t \Rightarrow \frac{dy}{dx} = \frac{2\cos t}{-2\sin t} = -\cot t.\ \therefore \frac{dy}{dx}\left(\frac{\pi}{4}\right) = -\cot\frac{\pi}{4} = -1.$

$\therefore$ Tangent line is $y - \sqrt{2} = -1\left(x - \sqrt{2}\right) \Rightarrow y = -x + 2\sqrt{2}.$

$\frac{dy'}{dt} = \csc^2 t \Rightarrow \frac{d^2y}{dx^2} = \frac{\csc^2 t}{-2\sin t} = -\frac{1}{2\sin^3 t} \Rightarrow \frac{d^2y}{dx^2}\left(\frac{\pi}{4}\right) = -\sqrt{2}$

$$\sqrt{2}\int_0 \sqrt{\frac{\sin^2 t}{1-\cos t}}\;dt =$$

$$\Big|_0^2 = 4. \qquad t^{-3/2} \Rightarrow \frac{d^2y}{dx^2}\left(\frac{1}{4}\right) = -2.$$

9. $t=-1 \Rightarrow x=5,\ y=1.\ \frac{dx}{dt}=4t,\ \frac{dy}{dt}=4t^3 \Rightarrow \frac{dy}{dx}=\frac{4t^3}{4t}=t^2 \Rightarrow \frac{dy}{dx}(-1)=(-1)^2=1.$ $\therefore$ Tangent line is $y-1=1(x-5)$ $\Rightarrow y=x-4.\ \frac{dy'}{dt}=2t \Rightarrow \frac{d^2y}{dx^2}=\frac{2t}{4t}=\frac{1}{2} \Rightarrow \frac{d^2y}{dx^2}(-1)=\frac{1}{2}$

13. $x^2-2tx+2t^2=4 \Rightarrow 2x\frac{dx}{dt}-2x-2t\frac{dx}{dt}+4t=0 \Rightarrow (2x-2t)\frac{dx}{dt}=2x-4t \Rightarrow \frac{dx}{dt}=\frac{2x-4t}{2x-2t}=\frac{x-2t}{x-t}.$ $2y^3-3t^2=4 \Rightarrow$ $6y^2\frac{dy}{dt}-6t=0 \Rightarrow \frac{dy}{dt}=\frac{6t}{6y^2}=\frac{t}{y^2}.\ \therefore \frac{dy}{dx}=\frac{dy/dt}{dx/dt}=\frac{t/y^2}{(x-2t)/(x-t)}=\frac{t(x-t)}{y^2(x-2t)}.$ $t=2 \Rightarrow x^2-2(2)x+2(2)^2=4$ $\Rightarrow x^2-4x+4=0 \Rightarrow (x-2)^2=0 \Rightarrow x=2.$ $t=2 \Rightarrow 2y^3-3(2)^2=4 \Rightarrow 2y^3=16 \Rightarrow y^3=8 \Rightarrow y=2.$ $\therefore \frac{dy}{dx}\Big|_{t=2}=\frac{2(2-2)}{(2)^2(2-2(2))}=0.$

15. $x+2x^{3/2}=t^2+t \Rightarrow \frac{dx}{dt}+3x^{1/2}\frac{dx}{dt}=2t+1 \Rightarrow \frac{dx}{dt}\left(1+3x^{1/2}\right)=2t+1 \Rightarrow \frac{dx}{dt}=\frac{2t+1}{1+3x^{1/2}}.$ $y\sqrt{t+1}+2t\sqrt{y}=4 \Rightarrow$ $\frac{dy}{dt}\sqrt{t+1}+y\left(\frac{1}{2}\right)(t+1)^{-1/2}+2\sqrt{y}+2t\left(\frac{1}{2}\right)y^{-1/2}\frac{dy}{dt}=0 \Rightarrow \frac{dy}{dt}\sqrt{t+1}+\frac{y}{2\sqrt{t+1}}+2\sqrt{y}+\frac{t}{\sqrt{y}}\frac{dy}{dt}=0 \Rightarrow$ $\Rightarrow \frac{dy}{dt}\left(\sqrt{t+1}+\frac{t}{\sqrt{y}}\right)=\frac{-y}{2\sqrt{t+1}}-2\sqrt{y} \Rightarrow \frac{dy}{dt}=\frac{\left(-y/\left(2\sqrt{t+1}\right)\right)-2\sqrt{y}}{\sqrt{t+1}+t/\sqrt{y}}=\frac{-y\sqrt{y}-4y\sqrt{t+1}}{2\sqrt{y}\,(t+1)+2t\sqrt{t+1}}$ $\therefore \frac{dy}{dx}=\frac{\left(-y\sqrt{y}-4y\sqrt{t+1}\right)/\left(2\sqrt{y}\,(t+1)+2t\sqrt{t+1}\right)}{(2t+1)/\left(1+3x^{1/2}\right)}.$ $t=0 \Rightarrow x+2x^{3/2}=0 \Rightarrow x\left(1+2x^{1/2}\right)=0 \Rightarrow x=0.$ $t=0$ $\Rightarrow y\sqrt{0+1}+2(0)\sqrt{y}=4 \Rightarrow y=4.\ \therefore \frac{dy}{dx}\Big|_{t=0}=\frac{\left(-4\sqrt{4}-4(4)\sqrt{0+1}\right)/\left(2\sqrt{4}\,(0+1)+2(0)\sqrt{0+1}\right)}{(2(0)+1)/\left(1+3(0)^{1/2}\right)}=-6.$

19. $\frac{dx}{dt}=t,\ \frac{dy}{dt}=(2t+1)^{1/2} \Rightarrow \sqrt{\left(\frac{dx}{dt}\right)^2+\left(\frac{dy}{dt}\right)^2}=\sqrt{t^2+\left((2t+1)^{1/2}\right)^2}=|t+1|=t+1$ since $0 \le t \le 4.$

$\therefore$ Length $=\displaystyle\int_0^4 (t+1)\,dt=\left[\frac{t^2}{2}+t\right]_0^4=12.$

23. $\frac{dx}{dt}=-\sin t,\ \frac{dy}{dt}=\cos t \Rightarrow \sqrt{\left(\frac{dx}{dt}\right)^2+\left(\frac{dy}{dt}\right)^2}=\sqrt{(-\sin t)^2+(\cos t)^2}=1.$

$\therefore$ Area $=\displaystyle\int_0^{2\pi} 2\pi(2+\sin t)1\,dt=2\pi[2t-\cos t]_0^{2\pi}=8\pi^2$

27. $\frac{dx}{dt}=2,\ \frac{dy}{dt}=1 \Rightarrow \sqrt{\left(\frac{dx}{dt}\right)^2+\left(\frac{dy}{dt}\right)^2}=\sqrt{2^2+1^2}=\sqrt{5}.$ $\therefore$ Area $=\displaystyle\int_0^1 2\pi(t+1)\sqrt{5}\,dt=$ $2\pi\sqrt{5}\left[\frac{t^2}{2}+t\right]_0^1=3\pi\sqrt{5}.$ The slant height is $\sqrt{5} \Rightarrow$ Area $=\pi(1+2)\sqrt{5}=3\pi\sqrt{5}.$

31. $x = x \Rightarrow \frac{dx}{dx} = 1$. $y = f(x) \Rightarrow \frac{dy}{dx} = f'(x)$. Then $L = \int_a^b \sqrt{\left(\frac{dx}{dt}\right)^2 + \left(\frac{dy}{dt}\right)^2}\, dt = \int_a^b \sqrt{\left(\frac{dx}{dx}\right)^2 + \left(\frac{dy}{dx}\right)^2}\, dx = \int_a^b \sqrt{1 + \left(\frac{dy}{dx}\right)^2}\, dx$.

35.

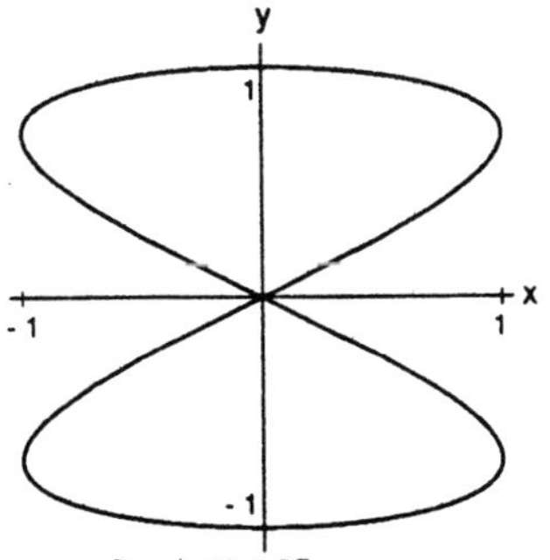

Graph 10.4.35

39.

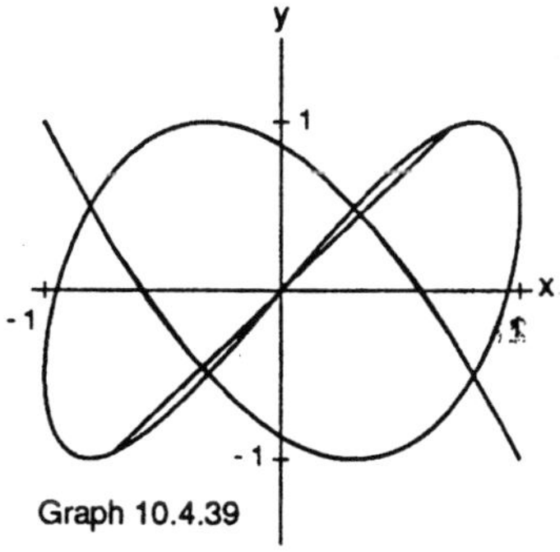

Graph 10.4.39

10.5 POLAR COORDINATES

1. a, c; b, d; e, k; g, j; h, f; i, l; m, o; n, p

5. a) $x = \sqrt{2}\cos\frac{\pi}{4} = 1,\ y = \sqrt{2}\sin\frac{\pi}{4} = 1 \Rightarrow (1,1)$

b) $x = 1\cos 0 = 1,\ y = 1\sin 0 = 1 \Rightarrow (1,0)$

c) $x = (0)\cos\frac{\pi}{2} = 0,\ y = 0\sin\frac{\pi}{2} = 0 \Rightarrow (0,0)$

d) $x = -\sqrt{2}\cos\frac{\pi}{4} = -1,\ y = -\sqrt{2}\sin\frac{\pi}{2} = -1 \Rightarrow (-1,-1)$

e) $x = -3\cos\frac{5\pi}{6} = \frac{3\sqrt{3}}{2},\ y = -3\sin\frac{5\pi}{6} = -\frac{3}{2} \Rightarrow \left(\frac{3\sqrt{3}}{2}, -\frac{3}{2}\right)$

f) $x = 5\cos(\tan^{-1}\frac{4}{3}) = 3,\ y = 5\sin(\tan^{-1}\frac{4}{3}) = 4 \Rightarrow (3,4)$

g) $x = -1\cos 7\pi = 1,\ y = -1\sin 7\pi = 0 \Rightarrow (1,0)$

h) $x = 2\sqrt{3}\cos\frac{2\pi}{3} = -\sqrt{3},\ y = 2\sqrt{3}\sin\frac{2\pi}{3} = 3 \Rightarrow (-\sqrt{3}, 3)$

9.

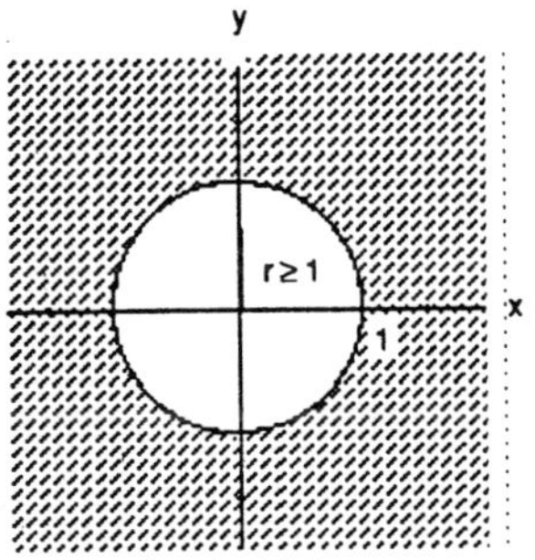

Graph 10.5.9

13.

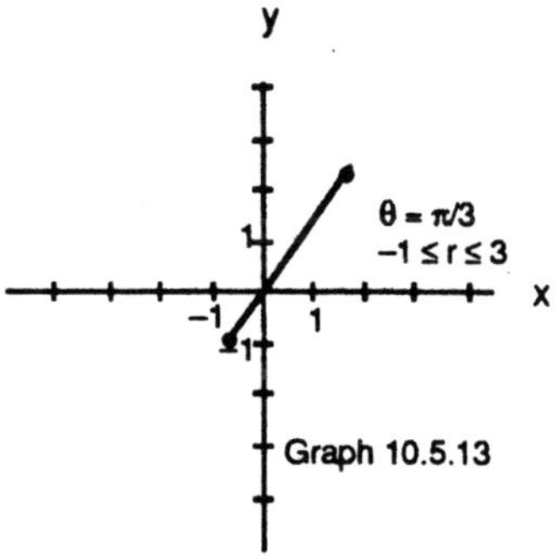

Graph 10.5.13

17.

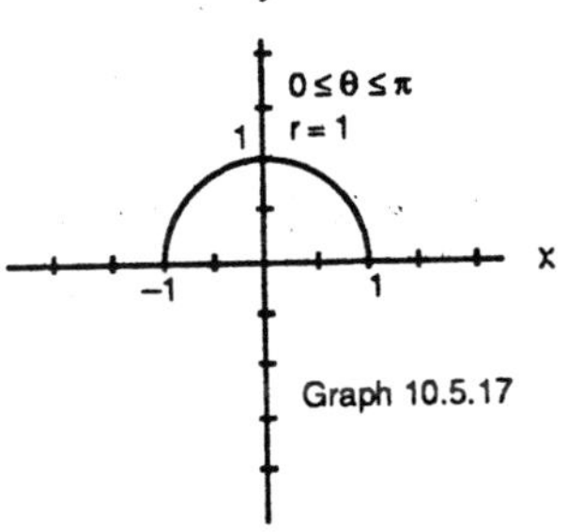

Graph 10.5.17

21.

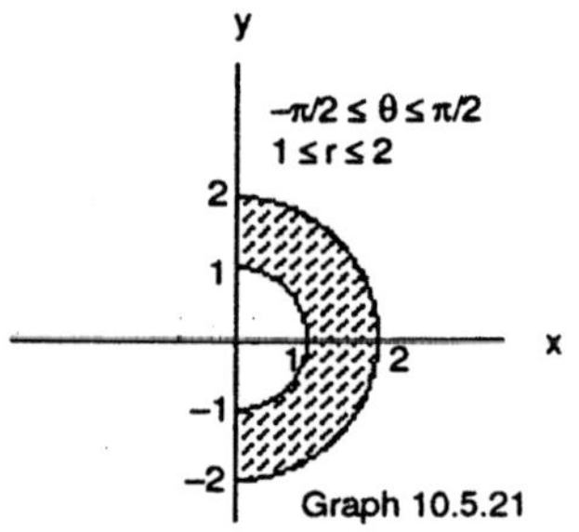

Graph 10.5.21

25. $r \sin \theta = 0 \Rightarrow y = 0$, the x–axis.

29. $r \cos \theta + r \sin \theta = 1 \Rightarrow x + y = 1$, line, $m = -1$, $b = 1$

33. $r = \dfrac{5}{\sin \theta - 2\cos \theta} \Rightarrow r \sin \theta - 2r \cos \theta = 5 \Rightarrow$

$y - 2x = 5$, line, $m = 2$, $b = 5$

37. $r = \csc \theta \, e^{r \cos \theta} \Rightarrow r \sin \theta = e^{r \cos \theta} \Rightarrow y = e^x$, the natural exponential function

41. $r^2 = -4 r \cos \theta \Rightarrow x^2 + y^2 = -4x \Rightarrow x^2 + 4x + y^2 = 0 \Rightarrow x^2 + 4x + 4 + y^2 = 4 \Rightarrow (x + 2)^2 + y^2 = 4$, a circle with center $(-2,0)$, radius $= 2$.

45. $r = 2 \cos \theta + 2 \sin \theta \Rightarrow r^2 = 2r \cos \theta + 2r \sin \theta \Rightarrow x^2 + y^2 = 2x + 2y \Rightarrow x^2 - 2x + y^2 - 2y = 0 \Rightarrow (x - 1)^2 + (y - 1)^2 = 2$, circle, center is $(1,1)$, $r = \sqrt{2}$.

49. $x = y \Rightarrow r \cos \theta = r \sin \theta \Rightarrow \theta = \frac{\pi}{4}$

53. $\frac{x^2}{9} + \frac{y^2}{4} = 1 \Rightarrow 4x^2 + 9y^2 = 36 \Rightarrow$

$4r^2\cos^2\theta + 9r^2\sin^2\theta = 36$

57. $x^2 + (y - 2)^2 = 4 \Rightarrow x^2 + y^2 - 4y + 4 = 4 \Rightarrow x^2 + y^2 = 4y \Rightarrow r^2 = 4 r \sin \theta \Rightarrow r = 4 \sin \theta$

61. a) $x = a \Rightarrow r \cos \theta = a \Rightarrow r = \dfrac{a}{\cos \theta} \Rightarrow r = a \sec \theta$ b) $y = b \Rightarrow r \sin \theta = b \Rightarrow r = \dfrac{b}{\sin \theta} \Rightarrow r = b \csc \theta$

10.6 POLAR GRAPHS

1.

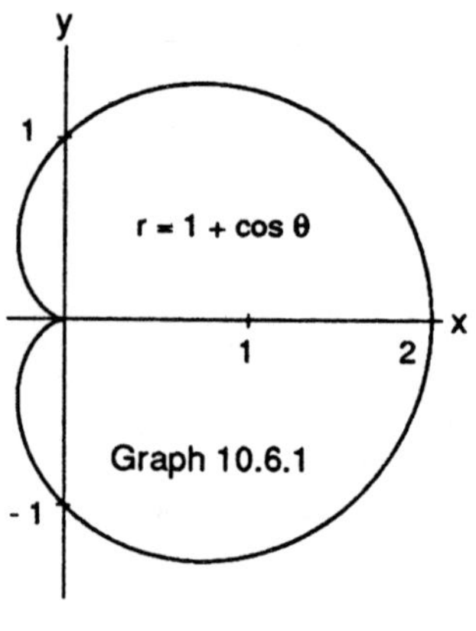

Graph 10.6.1

$1 + \cos(-\theta) = 1 + \cos \theta = r \Rightarrow$ symmetric about the x-axis.

$1 + \cos(-\theta) \ne -r$ and $1 + \cos(\pi - \theta) = 1 - \cos \theta \ne r \Rightarrow$ not symmetric about the y-axis. $\therefore$ not symmetric about the origin.

5.

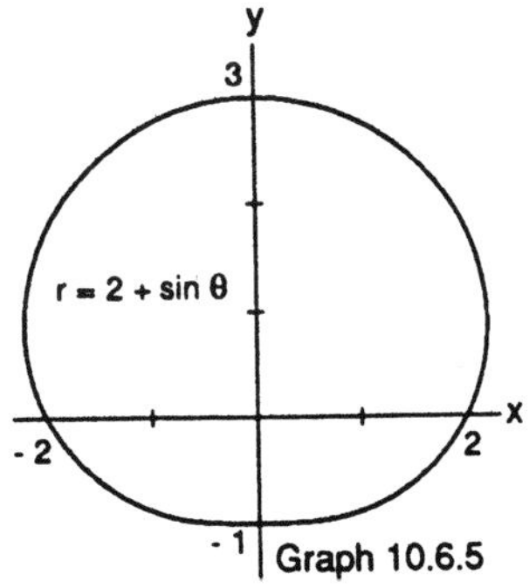

Graph 10.6.5

$2 + \sin(-\theta) = 2 - \sin\theta \neq r$ and $2 + \sin(\pi - \theta) = 2 + \sin\theta \neq -r \Rightarrow$ not symmetric about the x-axis. $2 + \sin(\pi - \theta) = 2 + \sin\theta = r \Rightarrow$ symmetric about the y-axis. $\therefore$ not symmetric about the origin.

9.

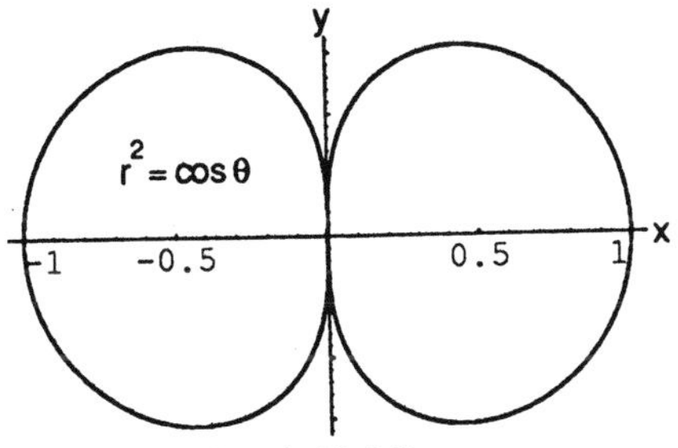

Graph 10.6.9

$\cos(-\theta) = \cos\theta = r^2 \Rightarrow (r,-\theta)$ and $(-r,-\theta)$ are on the graph when (r,θ) is on the graph $\Rightarrow$ symmetric about the x-axis and the y-axis. $\therefore$ symmetric about the origin.

17. $\theta = \frac{\pi}{4} \Rightarrow r = 1 \Rightarrow \left(1, \frac{\pi}{4}\right)$. $\theta = -\frac{\pi}{4} \Rightarrow r = -1 \Rightarrow \left(-1, -\frac{\pi}{4}\right)$

$\theta = \frac{3\pi}{4} \Rightarrow r = -1 \Rightarrow \left(-1, \frac{3\pi}{4}\right)$. $\theta = -\frac{3\pi}{4} \Rightarrow r = 1 \Rightarrow \left(1, -\frac{3\pi}{4}\right)$

$r' = \frac{dr}{d\theta} = 2\cos 2\theta \Rightarrow$

$$\text{Slope} = \frac{r'\sin\theta + r\cos\theta}{r'\cos\theta - r\sin\theta} = \frac{2\cos 2\theta\sin\theta + r\cos\theta}{2\cos 2\theta\cos\theta - r\sin\theta} \Rightarrow$$

$$\text{Slope at } \left(1, \frac{\pi}{4}\right) = \frac{2\cos\left(\frac{\pi}{2}\right)\sin\frac{\pi}{4} + (1)\cos\frac{\pi}{4}}{2\cos\left(\frac{\pi}{2}\right)\cos\frac{\pi}{4} - (1)\sin\frac{\pi}{4}} = -1$$

$$\text{Slope at } \left(-1, -\frac{\pi}{4}\right) = \frac{2\cos\left(-\frac{\pi}{2}\right)\sin\left(-\frac{\pi}{4}\right) + (-1)\cos\left(-\frac{\pi}{4}\right)}{2\cos\left(-\frac{\pi}{2}\right)\cos\left(-\frac{\pi}{4}\right) - (-1)\sin\left(-\frac{\pi}{4}\right)} = 1$$

$$\text{Slope at } \left(-1, \frac{3\pi}{4}\right) = \frac{2\cos\left(\frac{3\pi}{2}\right)\sin\left(\frac{3\pi}{4}\right) + (-1)\cos\left(\frac{3\pi}{4}\right)}{2\cos\left(\frac{3\pi}{2}\right)\cos\left(\frac{3\pi}{4}\right) - (-1)\sin\left(\frac{3\pi}{4}\right)} = 1$$

$$\text{Slope at } \left(1, -\frac{3\pi}{4}\right) = \frac{2\cos\left(-\frac{3\pi}{2}\right)\sin\left(-\frac{3\pi}{4}\right) + (1)\cos\left(-\frac{3\pi}{4}\right)}{2\cos\left(-\frac{3\pi}{2}\right)\cos\left(-\frac{3\pi}{4}\right) - (1)\sin\left(-\frac{3\pi}{4}\right)} = -1$$

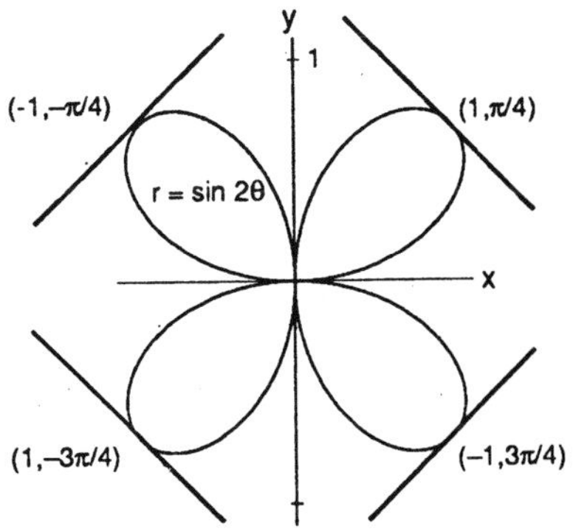

Graph 10.6.17

21.

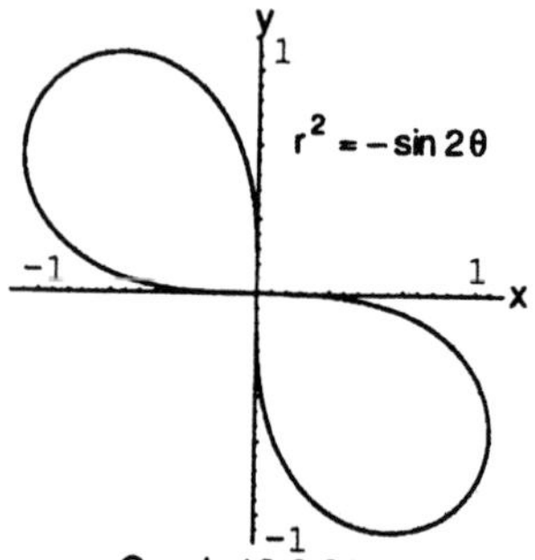

Graph 10.6.21

Since (r,θ) on the graph $\Rightarrow$ $(-r,\theta)$ is on the graph $\left((\pm r)^2 = -\sin 2\theta \Rightarrow r^2 = -\sin 2\theta\right)$, the graph is symmetric about the origin. But $-\sin 2(-\theta) = -(-\sin 2\theta) = \sin 2\theta \neq r^2$ and $-\sin 2(\pi - \theta) = -\sin(2\pi - 2\theta) = -\sin(-2\theta) = -(-\sin 2\theta) = \sin 2\theta \neq r^2 \Rightarrow$ the graph is not symmetric about the x-axis. $\therefore$ the graph is not symmetric about the y-axis.

25. a)

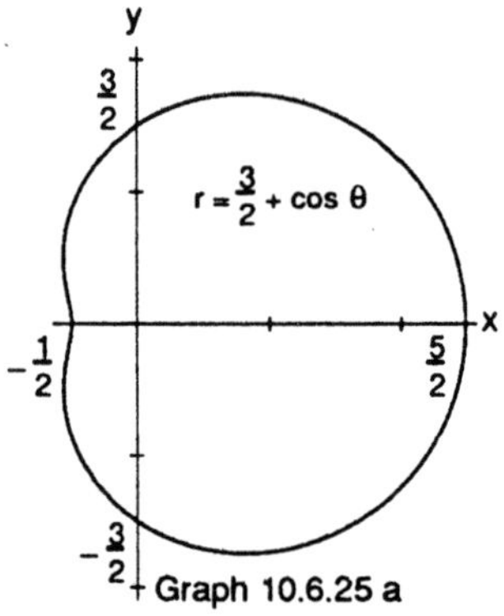

Graph 10.6.25 a

25. b)

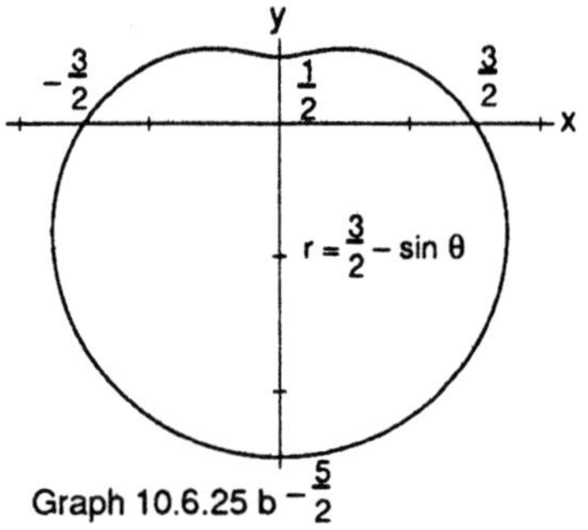

Graph 10.6.25 b

29. $\left(2, \frac{3\pi}{4}\right)$ is the same point as $\left(-2, -\frac{\pi}{4}\right)$. $r = 2\sin 2\left(-\frac{\pi}{4}\right) = 2\sin\left(-\frac{\pi}{2}\right) = -2 \Rightarrow \left(-2, -\frac{\pi}{4}\right)$ is on the graph $\Rightarrow \left(2, \frac{3\pi}{4}\right)$ is on the graph.

33. $2\sin\theta = 2\sin 2\theta \Rightarrow \sin\theta = \sin 2\theta \Rightarrow \sin\theta = 2\sin\theta\cos\theta$
$\Rightarrow \sin\theta - 2\sin\theta\cos\theta = 0 \Rightarrow \sin\theta(1 - 2\cos\theta) = 0 \Rightarrow$
$\sin\theta = 0$ or $\cos\theta = 1/2 \Rightarrow \theta = 0, \frac{\pi}{3}$, or $-\frac{\pi}{3}$.
$\theta = 0 \Rightarrow r = 0$, $\theta = \frac{\pi}{3} \Rightarrow r = \sqrt{3}$, $\theta = -\frac{\pi}{3} \Rightarrow r = -\sqrt{3}$.
The points of intersection are (0,0), $\left(\sqrt{3}, \frac{\pi}{3}\right)$, $\left(-\sqrt{3}, -\frac{\pi}{3}\right)$.

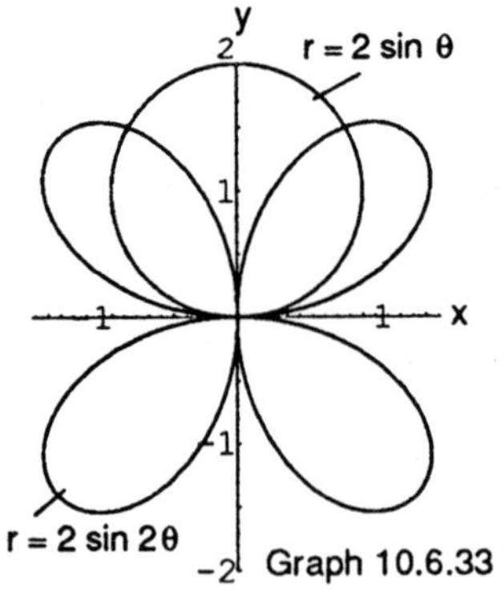

Graph 10.6.33

37. $1 = 2\sin 2\theta \Rightarrow \sin 2\theta = \frac{1}{2} \Rightarrow 2\theta = \frac{\pi}{6}, \frac{5\pi}{6}, \frac{13\pi}{6}, \frac{17\pi}{6} \Rightarrow$
$\theta = \frac{\pi}{12}, \frac{5\pi}{12}, \frac{13\pi}{12}, \frac{17\pi}{12}$. The points of intersection are
$\left(1, \frac{\pi}{12}\right), \left(1, \frac{5\pi}{12}\right), \left(1, \frac{13\pi}{12}\right), \left(1, \frac{17\pi}{12}\right)$.
No other points are found by graphing.

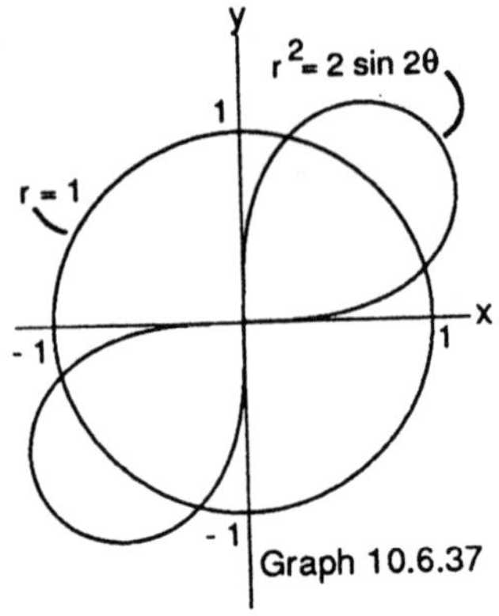

Graph 10.6.37

39. a) $r^2 = -4\cos\theta \Rightarrow \cos\theta = -\frac{r^2}{4}$. $r = 1 - \cos\theta \Rightarrow r = 1 - \left(-\frac{r^2}{4}\right) \Rightarrow 0 = r^2 - 4r + 4 \Rightarrow (r - 2)^2 = 0 \Rightarrow$

$r = 2$. $\therefore \cos\theta = -\frac{2^2}{4} = -1 \Rightarrow \theta = \pi$ $\therefore (2,\pi)$ is a point of intersection.

b) $r = 0 \Rightarrow 0^2 = 4\cos\theta \Rightarrow \cos\theta = 0 \Rightarrow \theta = \frac{\pi}{2}, \frac{3\pi}{2} \Rightarrow \left(0,\frac{\pi}{2}\right)$ or $\left(0,\frac{3\pi}{2}\right)$ is on the graph.

$r = 0 \Rightarrow 0 = 1 - \cos\theta \Rightarrow \cos\theta = 1 \Rightarrow \theta = 0 \Rightarrow (0,0)$ is on the graph. Since $(0,0) = \left(0,\frac{\pi}{2}\right)$,

the graphs intersect at the origin.

43. $1 = 2\sin 2\theta \Rightarrow \sin 2\theta = \frac{1}{2} \Rightarrow 2\theta = \frac{\pi}{6}, \frac{5\pi}{6}, \frac{13\pi}{6}, \frac{17\pi}{6}$

$\Rightarrow \theta = \frac{\pi}{12}, \frac{5\pi}{12}, \frac{13\pi}{12}, \frac{17\pi}{12}$. Points of intersection are

$\left(1,\frac{\pi}{12}\right), \left(1,\frac{5\pi}{12}\right), \left(1,\frac{13\pi}{12}\right)$, and $\left(1,\frac{17\pi}{12}\right)$.

Points of intersection $\left(1,\frac{7\pi}{12}\right), \left(1,\frac{11\pi}{12}\right), \left(1,\frac{19\pi}{12}\right)$, and

$\left(1,\frac{23\pi}{12}\right)$ found by graphing and symmetry.

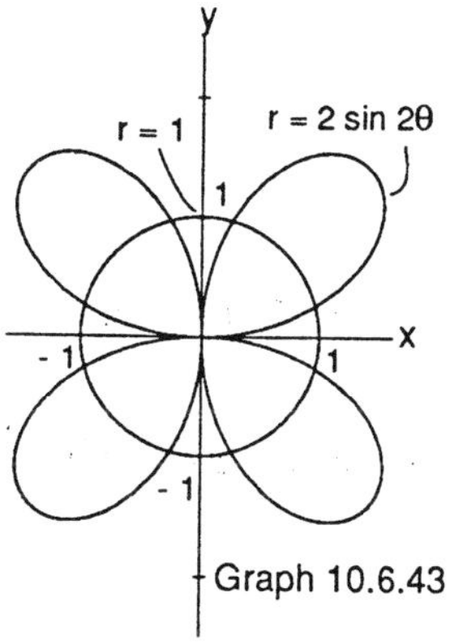

Graph 10.6.43

47. a)

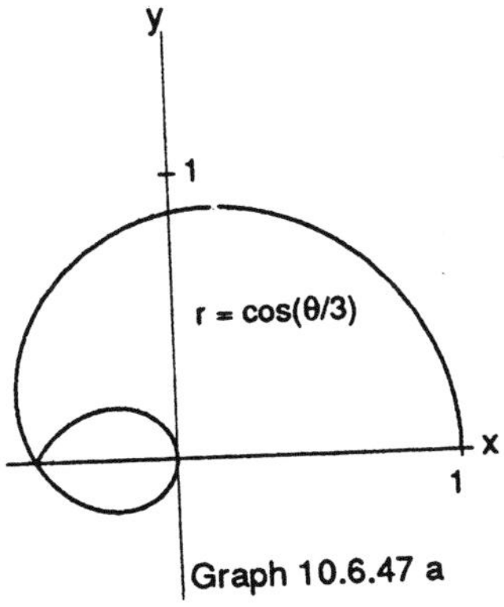

Graph 10.6.47 a

47. b)

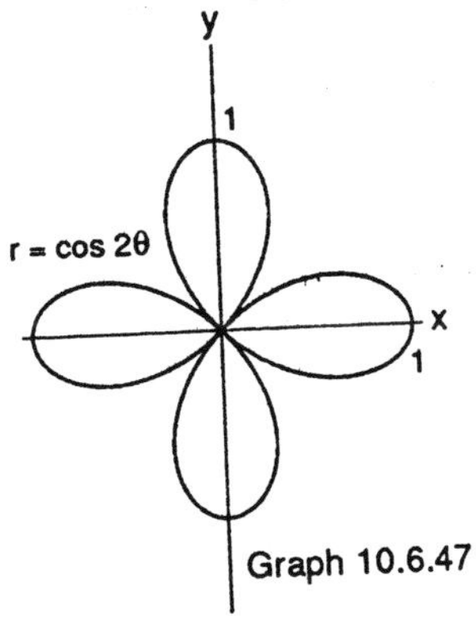

Graph 10.6.47 b

47. c)

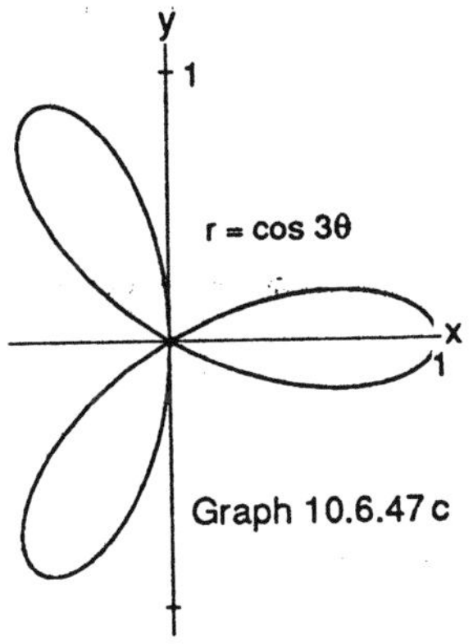

Graph 10.6.47 c

47. d)

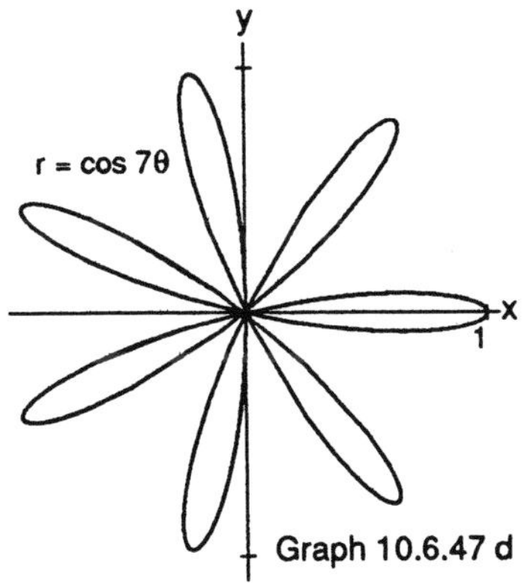

Graph 10.6.47 d

10.7 POLAR EQUATIONS FOR CONIC SECTIONS

1. $r\cos\left(\theta - \frac{\pi}{6}\right) = 5 \Rightarrow r\left(\cos\theta\cos\frac{\pi}{6} + \sin\theta\sin\frac{\pi}{6}\right) = 5 \Rightarrow \frac{\sqrt{3}}{2}r\cos\theta + \frac{1}{2}r\sin\theta = 5 \Rightarrow$
$\frac{\sqrt{3}}{2}x + \frac{1}{2}y = 5 \Rightarrow \sqrt{3}x + y = 10.$

5. $r\cos\left(\theta - \frac{\pi}{4}\right) = \sqrt{2} \Rightarrow r\left(\cos\theta\cos\frac{\pi}{4} + \sin\theta\sin\frac{\pi}{4}\right) =$
$\sqrt{2} \Rightarrow \frac{1}{\sqrt{2}}r\cos\theta + \frac{1}{\sqrt{2}}r\sin\theta = \sqrt{2} \Rightarrow$
$\frac{1}{\sqrt{2}}x + \frac{1}{\sqrt{2}}y = \sqrt{2} \Rightarrow x + y = 2.$

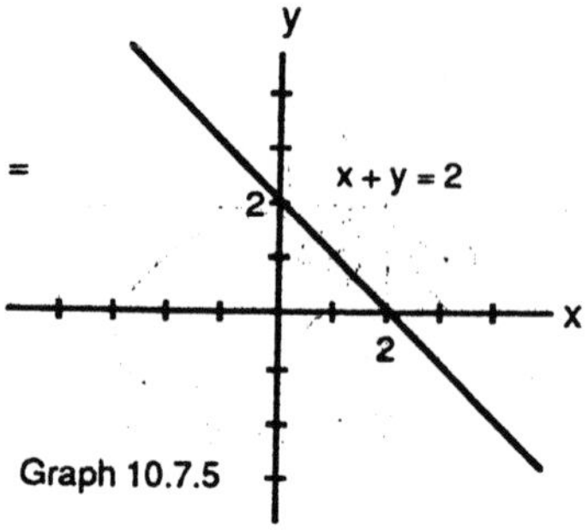

Graph 10.7.5

9. $r = 2(4)\cos\theta = 8\cos\theta$

13.

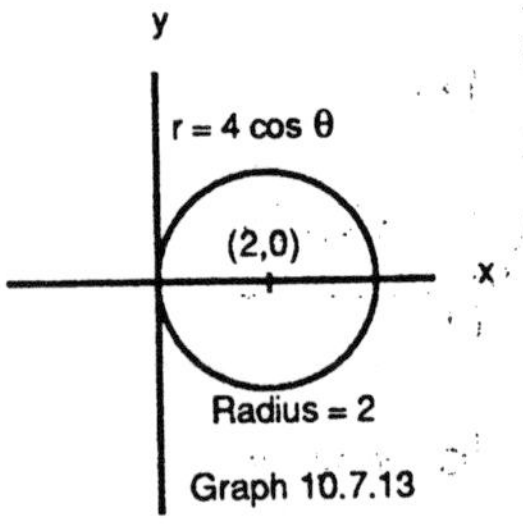

Graph 10.7.13

17. $(x-6)^2 + y^2 = 36 \Rightarrow C = (6,0),\ a = 6 \Rightarrow r = 12\cos\theta$ is the polar equation.

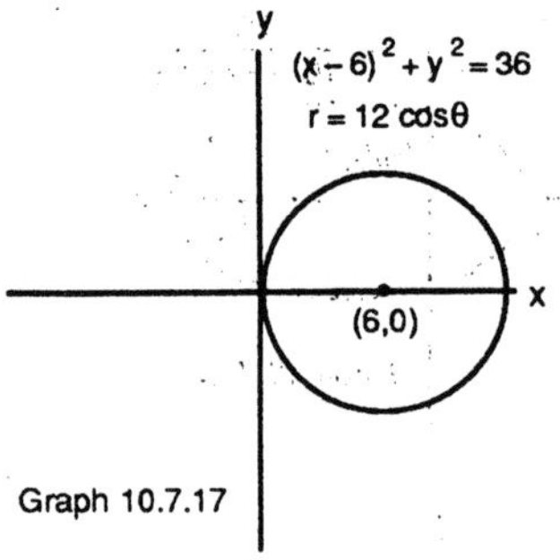

Graph 10.7.17

21. $x^2 + 2x + y^2 = 0 \Rightarrow (x+1)^2 + y^2 = 1 \Rightarrow C = (-1,0),\ a = 1$
$\Rightarrow r = -2\cos\theta$ is the polar equation.

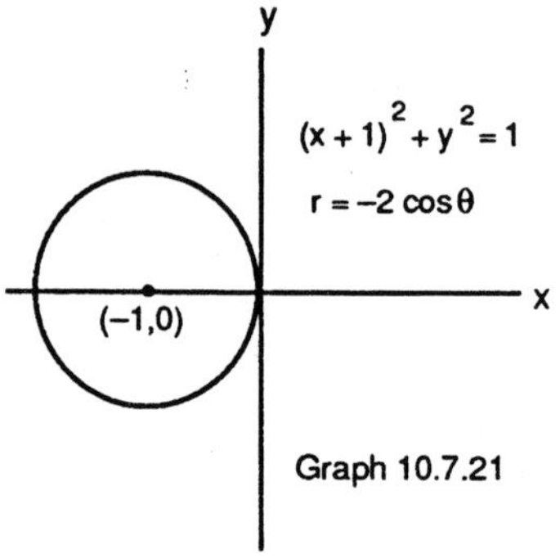

Graph 10.7.21

25. $e = 1, x = 2 \Rightarrow k = 2 \Rightarrow r = \frac{2(1)}{1 + (1)\cos\theta} = \frac{2}{1 + \cos\theta}$

29. $e = \frac{1}{2}, x = 1 \Rightarrow k = 1 \Rightarrow r = \frac{(1/2)(1)}{1 + (1/2)\cos\theta} = \frac{1}{2 + \cos\theta}$

33. $r = \frac{1}{1 + \cos\theta} \Rightarrow e = 1, k = 1 \Rightarrow x = 1$

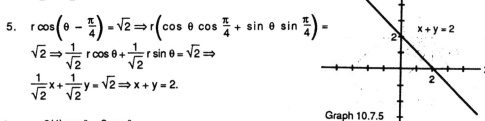

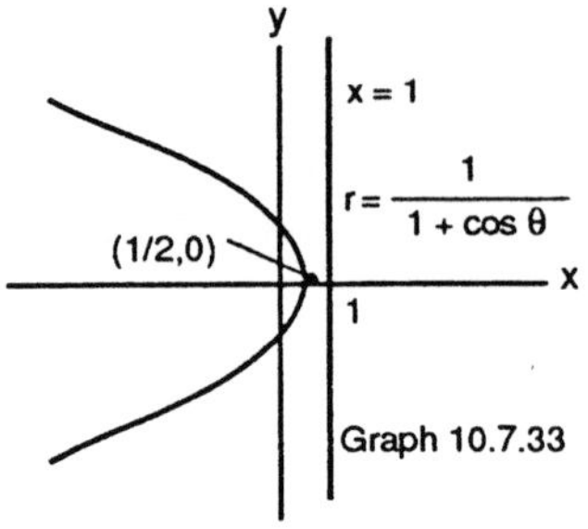

Graph 10.7.33

37. $r = \dfrac{400}{16 + 8\sin\theta} \Rightarrow r = \dfrac{\frac{400}{16}}{1 + \frac{8}{16}\sin\theta} \Rightarrow r = \dfrac{25}{1 + \frac{1}{2}\sin\theta} \Rightarrow$

$e = \frac{1}{2}$, $k = 50$. $a(1 - e^2) = ke \Rightarrow a\left(1 - \left(\frac{1}{2}\right)^2\right) = 25 \Rightarrow$

$\frac{3}{4}a = 25 \Rightarrow a = \frac{100}{3}$. $a - ae = \frac{100}{3} - \frac{100}{3}\left(\frac{1}{2}\right) = \frac{50}{3}$.

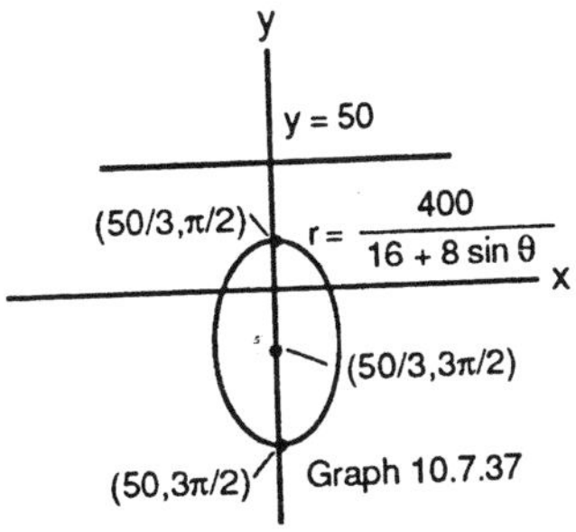

Graph 10.7.37

41.

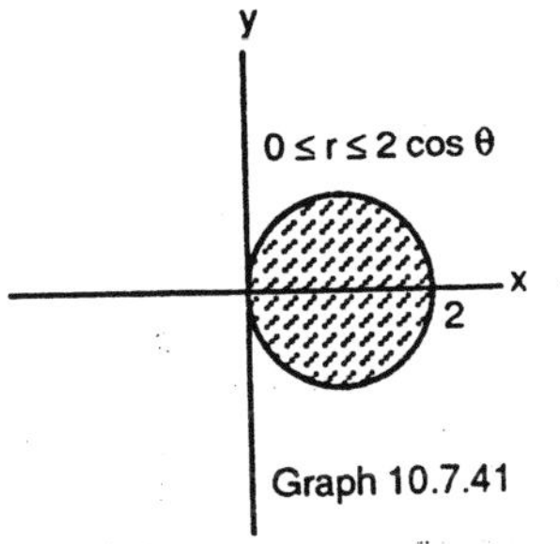

Graph 10.7.41

45. a) $r = 4\sin\theta \Rightarrow r^2 = 4r\sin\theta \Rightarrow x^2 + y^2 = 4y$ and $r = \sqrt{3}\sec\theta$

$\Rightarrow r = \dfrac{\sqrt{3}}{\cos\theta} \Rightarrow r\cos\theta = \sqrt{3} \Rightarrow x = \sqrt{3}$. $x = \sqrt{3} \Rightarrow$

$\left(\sqrt{3}\right)^2 + y^2 = 4y \Rightarrow y^2 - 4y + 3 = 0 \Rightarrow (y-3)(y-1) = 0 \Rightarrow$

$y = 3$ or $y = 1$. ∴ in Cartesian coordinates, the points are

$\left(\sqrt{3}, 3\right), \left(\sqrt{3}, 1\right)$. In polar coordinates, $4\sin\theta = \sqrt{3}\sec\theta$

$\Rightarrow 4\sin\theta\cos\theta = \sqrt{3} \Rightarrow 2\sin\theta\cos\theta = \frac{\sqrt{3}}{2} \Rightarrow \sin 2\theta = \frac{\sqrt{3}}{2} \Rightarrow$

$2\theta = \frac{\pi}{3}$ or $\frac{2\pi}{3} \Rightarrow \theta = \frac{\pi}{6}$ or $\frac{\pi}{3}$. $\theta = \frac{\pi}{6} \Rightarrow r = 2$, $\theta = \frac{\pi}{3} \Rightarrow r = 2\sqrt{3}$

$\Rightarrow \left(2, \frac{\pi}{6}\right)$ and $\left(2\sqrt{3}, \frac{\pi}{3}\right)$ are the points in polar coordinates.

b)

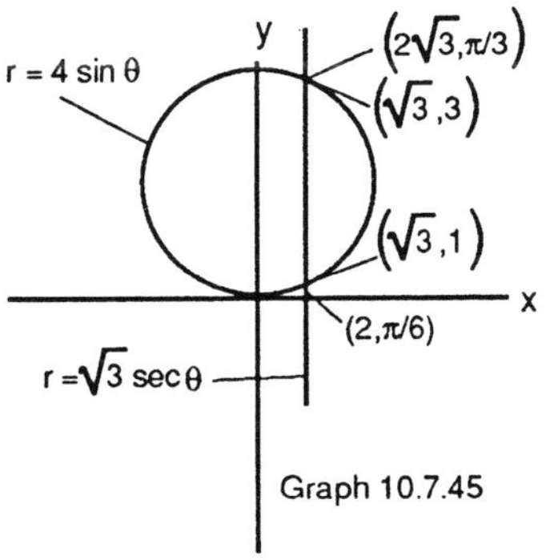

Graph 10.7.45

49. a)

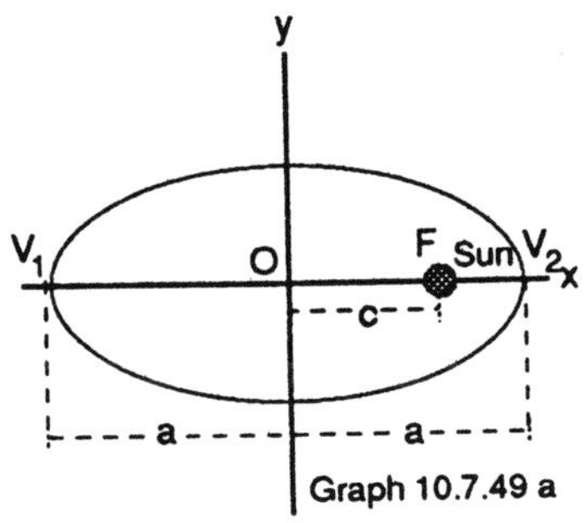

Graph 10.7.49 a

Let the ellipse be the orbit, with the Sun at one focus. Then $r_{max} =$

$a + c$, $r_{min} = a - c \Rightarrow \dfrac{r_{max} - r_{min}}{r_{max} + r_{min}} = \dfrac{(a + c) - (a - c)}{(a + c) + (a - c)} = \dfrac{2c}{2a} = \dfrac{c}{a} = e$.

b) Let F_1, F_2 be the foci. Then $PF_1 + PF_2 = 10$ where P is any point on the ellipse. If P is a vertex, then $PF_1 = a + c$, $PF_2 = a - c \Rightarrow (a + c) + (a - c) = 10 \Rightarrow$ $2a = 10 \Rightarrow a = 5$. Since $e = \frac{c}{a}$, $0.2 = \frac{c}{5} \Rightarrow c = 1.0 \Rightarrow$ the pins are 2 inches apart.

53.

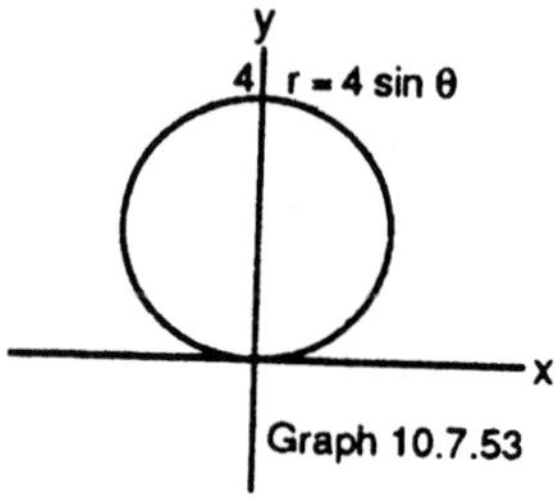

Graph 10.7.53

57.

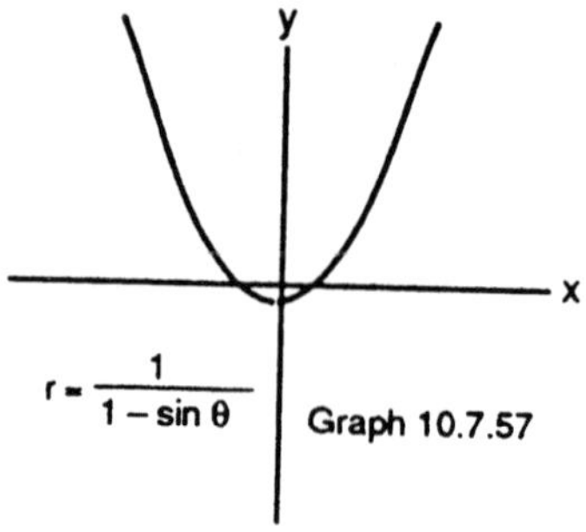

Graph 10.7.57

10.8 INTEGRATION IN POLAR COORDINATES

1. $A = \int_0^{\pi/4} \frac{1}{2}\left(2\sqrt{\cos\theta}\right)^2 d\theta = \int_0^{\pi/4} 2\cos\theta\, d\theta = [2\sin\theta]_0^{\pi/4} = 2\sin\frac{\pi}{4} - 2\sin 0 = \sqrt{2}$

5. $A = \int_0^{2\pi} \frac{1}{2}(4 + 2\cos\theta)^2 d\theta = \int_0^{2\pi} \frac{1}{2}(16 + 16\cos\theta + 4\cos^2\theta)\, d\theta = \int_0^{2\pi} \left(8 + 8\cos\theta + 2\left(\frac{1 + \cos 2\theta}{2}\right)\right) d\theta$

$= \int_0^{2\pi} (9 + 8\cos\theta + \cos 2\theta)\, d\theta = \left[9\theta + 8\sin\theta + \frac{1}{2}\sin 2\theta\right]_0^{2\pi} = 18\pi$

9. $A = \int_0^{\pi/2} \frac{1}{2}(4\sin 2\theta)\, d\theta = \int_0^{\pi/2} 2\sin 2\theta\, d\theta = [-\cos 2\theta]_0^{\pi/2} = 2$

13. $r = 2, r = 2(1 - \cos\theta) \Rightarrow 2 = 2(1 - \cos\theta) \Rightarrow \cos\theta = 0 \Rightarrow \theta = \pm\frac{\pi}{2}$. Sketch a graph to see the region.

$A = 2\int_0^{\pi/2} \frac{1}{2}(2(1 - \cos\theta))^2 d\theta + \frac{1}{2}$ of the area of the circle $= \int_0^{\pi/2} 4(1 - 2\cos\theta + \cos^2\theta)d\theta + \frac{1}{2}\pi(2)^2$

$= \int_0^{\pi/2} \left(4 - 8\cos\theta + 4\left(\frac{1 + \cos 2\theta}{2}\right)\right)d\theta + 2\pi = \int_0^{\pi/2} (6 - 8\cos\theta + 2\cos 2\theta)d\theta + 2\pi =$

$[6\theta - 8\sin\theta + \sin 2\theta]_0^{\pi/2} + 2\pi = 5\pi - 8$

17. $r = 1, r = -2\cos\theta \Rightarrow 1 = -2\cos\theta \Rightarrow \cos\theta = -\frac{1}{2} \Rightarrow \theta = \frac{2\pi}{3}$ in quadrant II.

$A = 2\int_{2\pi/3}^{\pi} \frac{1}{2}\left((-2\cos\theta)^2 - 1^2\right)d\theta = \int_{2\pi/3}^{\pi} \left(4\cos^2\theta - 1\right)d\theta = \int_{2\pi/3}^{\pi} (2(1 + \cos 2\theta) - 1)d\theta$

$= \int_{2\pi/3}^{\pi} (1 + 2\cos 2\theta)d\theta = [\theta + \sin 2\theta]_{2\pi/3}^{\pi} = \frac{\pi}{3} + \frac{\sqrt{3}}{2}$

21. a)

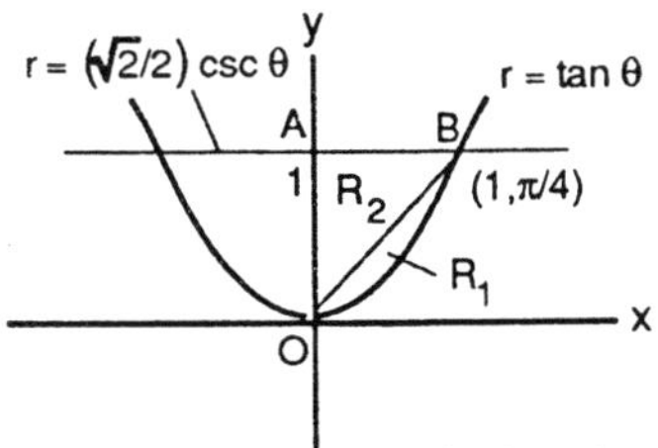

Graph 10.8.21 a

$r = \tan\theta$, $r = (\sqrt{2}/2)\csc\theta \Rightarrow \tan\theta = (\sqrt{2}/2)\csc\theta \Rightarrow \sin^2\theta = (\sqrt{2}/2)\cos\theta \Rightarrow 1 - \cos^2\theta = (\sqrt{2}/2)\cos\theta \Rightarrow \cos^2\theta + (\sqrt{2}/2)\cos\theta - 1 = 0 \Rightarrow \cos\theta = -\sqrt{2}$ or $\frac{\sqrt{2}}{2}$ (Use the quadratic formula.) $\Rightarrow \theta = \frac{\pi}{4}$ (the solution in the first quadrant).

$\therefore$ The area of R_1 is $A_1 = \int_0^{\pi/4} \frac{1}{2}\tan^2\theta\, d\theta = \frac{1}{2}\int_0^{\pi/4}(\sec^2\theta - 1)\, d\theta = \frac{1}{2}[\tan\theta - \theta]_0^{\pi/4} = \left(\frac{1}{2}\left(\tan\frac{\pi}{4} - \frac{\pi}{4}\right)\right) = \frac{1}{2} - \frac{\pi}{8}.$

$AO = (\sqrt{2}/2)\csc\frac{\pi}{2} = \frac{\sqrt{2}}{2}$, $OB = (\sqrt{2}/2)\csc\frac{\pi}{4} = 1$

$\Rightarrow AB = \sqrt{1^2 - (\sqrt{2}/2)^2} = \sqrt{2}/2 \Rightarrow$ the area of R_2 is $A_2 = \frac{1}{2}(\sqrt{2}/2)(\sqrt{2}/2) = 1/4$. $\therefore$ the area of the region shaded in the text is $2\left(\frac{1}{2} - \frac{\pi}{8} + \frac{1}{4}\right) = \frac{3}{2} - \frac{\pi}{4}$. Note: the area must be found this way since no common interval generates the region. For example, the interval $0 \le \theta \le \pi/4$ generates the arc OB of $r = \tan\theta$ but does not generate the segment AB of the line $r = \csc\theta$. Instead the interval generates the half–line from B to $+\infty$ on the line $r = \csc\theta$.

b) $\lim_{\theta \to (\pi/2)^-} \tan\theta = +\infty$. The line $x = 1$ is $r = \sec\theta$ in polar coordinates. Then $\lim_{\theta \to (\pi/2)^-} (\tan\theta - \sec\theta) = \lim_{\theta \to (\pi/2)^-}\left(\frac{\sin\theta}{\cos\theta} - \frac{1}{\cos\theta}\right) = \lim_{\theta \to (\pi/2)^-}\left(\frac{\sin\theta - 1}{\cos\theta}\right) = \lim_{\theta \to (\pi/2)^-}\left(\frac{\cos\theta}{-\sin\theta}\right) = 0 \Rightarrow r = \tan\theta \to r = \sec\theta$ as $\theta \to \frac{\pi}{2}^- \Rightarrow r = \sec\theta$ ($x = 1$) is a vertical asymptote of $r = \tan\theta$. Similarly, $r = -\sec\theta$ is the polar equation of $x = -1$ and $\lim_{\theta \to (-\pi/2)^+}(\tan\theta - (-\sec\theta)) = 0 \Rightarrow r = -\sec\theta$ ($x = -1$) is a vertical asymptote of $r = \tan\theta$.

25. $r = 1 + \cos\theta \Rightarrow \frac{dr}{d\theta} = -\sin\theta$. $\therefore$ Length $= \int_0^{2\pi}\sqrt{(1 + \cos\theta)^2 + (-\sin\theta)^2}\, d\theta = 2\int_0^{\pi}\sqrt{1 + 2\cos\theta + \cos^2\theta + \sin^2\theta}\, d\theta = 2\int_0^{\pi}\sqrt{2 + 2\cos\theta}\, d\theta = 2\int_0^{\pi}\sqrt{\frac{4(1 + \cos\theta)}{2}}\, d\theta = 4\int_0^{\pi}\sqrt{\frac{1 + \cos\theta}{2}}\, d\theta = 4\int_0^{\pi}\cos\frac{1}{2}\theta\, d\theta = 4\left[2\sin\frac{1}{2}\theta\right]_0^{\pi} = 8$

29. $r = \sqrt{1 + \cos 2\theta} \Rightarrow \frac{dr}{d\theta} = \frac{1}{2}(1 + \cos 2\theta)^{-1/2}(-2\sin 2\theta) = \frac{-\sin 2\theta}{\sqrt{1 + \cos 2\theta}} \Rightarrow \left(\frac{dr}{d\theta}\right)^2 = \frac{\sin^2 2\theta}{1 + \cos 2\theta}$.

$r^2 = 1 + \cos 2\theta \Rightarrow r^2 + \left(\frac{dr}{d\theta}\right)^2 = 1 + \cos 2\theta + \frac{\sin^2 2\theta}{1 + \cos 2\theta} = 1 + \cos 2\theta + \left(\frac{\sin^2 2\theta}{1 + \cos 2\theta}\right)\left(\frac{1 - \cos 2\theta}{1 - \cos 2\theta}\right) = 1 + \cos 2\theta + \frac{\sin^2 2\theta\,(1 - \cos 2\theta)}{1 - \cos^2 2\theta} = 1 + \cos 2\theta + \frac{\sin^2 2\theta\,(1 - \cos 2\theta)}{\sin^2 2\theta} = 1 + \cos 2\theta + 1 - \cos 2\theta = 2.$

$\therefore L = \int_0^{\pi\sqrt{2}}\sqrt{2}\, d\theta = [\sqrt{2}\,\theta]_0^{\pi\sqrt{2}} = 2\pi.$

33. $r^2 = \cos 2\theta \Rightarrow r = \pm\sqrt{\cos 2\theta}$. Use $r = \sqrt{\cos 2\theta}$ on $\left[0, \frac{\pi}{4}\right]$. Then $\frac{dr}{d\theta} = \frac{1}{2}(\cos 2\theta)^{-1/2}(-\sin 2\theta)(2) = \frac{-\sin 2\theta}{\sqrt{\cos 2\theta}}$. $\therefore$ Surface Area $= \int_0^{\pi/4} 2\pi\sqrt{\cos 2\theta}\sin\theta\sqrt{\left(\sqrt{\cos 2\theta}\right)^2 + \left(\frac{-\sin 2\theta}{\sqrt{\cos 2\theta}}\right)^2}\,d\theta =$

$$\int_0^{\pi/4} 2\pi\sqrt{\cos 2\theta}\sin\theta\sqrt{\cos 2\theta + \frac{\sin^2 2\theta}{\cos 2\theta}}\,d\theta = \int_0^{\pi/4} 2\pi\sqrt{\cos 2\theta}\sin\theta\sqrt{\frac{1}{\cos 2\theta}}\,d\theta =$$

$$\int_0^{\pi/4} 2\pi\sin\theta\,d\theta = \left[-2\pi\cos\theta\right]_0^{\pi/4} = \pi\left(2 - \sqrt{2}\right)$$

37. a) $r = 2f(\theta),\ \alpha \le \theta \le \beta \Rightarrow \frac{dr}{d\theta} = 2f'(\theta) \Rightarrow r^2 + \left(\frac{dr}{d\theta}\right)^2 = (2f(\theta))^2 + (2f'(\theta))^2 = 4\left((f(\theta))^2 + (f'(\theta))^2\right) \Rightarrow$

Length $= \int_\alpha^\beta \sqrt{4\left((f(\theta))^2 + (f'(\theta))^2\right)}\,d\theta = 2\int_\alpha^\beta \sqrt{(f(\theta))^2 + (f'(\theta))^2}\,d\theta$ or twice the length of $r = f(\theta)$, $\alpha \le \theta \le \beta$.

b) Again $r = 2f(\theta) \Rightarrow r^2 + \left(\frac{dr}{d\theta}\right)^2 = (2f(\theta))^2 + (2f'(\theta))^2 = 4\left((f(\theta))^2 + (f'(\theta))^2\right) \Rightarrow$ Area $=$

$\int_\alpha^\beta 2\pi(2f(\theta))\sin\theta\sqrt{4\left((f(\theta))^2 + (f'(\theta))^2\right)}\,d\theta = 4\int_\alpha^\beta 2\pi f(\theta)\sqrt{(f(\theta))^2 + (f'(\theta))^2}\,d\theta$ or four times the area of the surface generated by revolving $r = f(\theta)$, $\alpha \le \theta \le \beta$, about the x–axis.

10.P PRACTICE EXERCISES

1. $x^2 = -4y \Rightarrow y = -\frac{x^2}{4} \Rightarrow 4p = 4 \Rightarrow p = 1$

$\therefore$ Focus: (0,–1); Directrix: $y = 1$

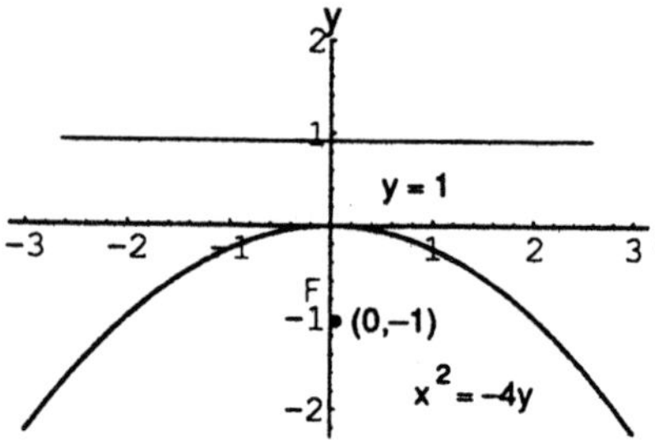

Graph 10.P.1

5. $16x^2 + 7y^2 = 112 \Rightarrow \frac{x^2}{7} + \frac{y^2}{16} = 1 \Rightarrow$

$c^2 = 16 - 7 = 9 \Rightarrow c = 3.\ e = \frac{c}{a} = \frac{3}{4}$

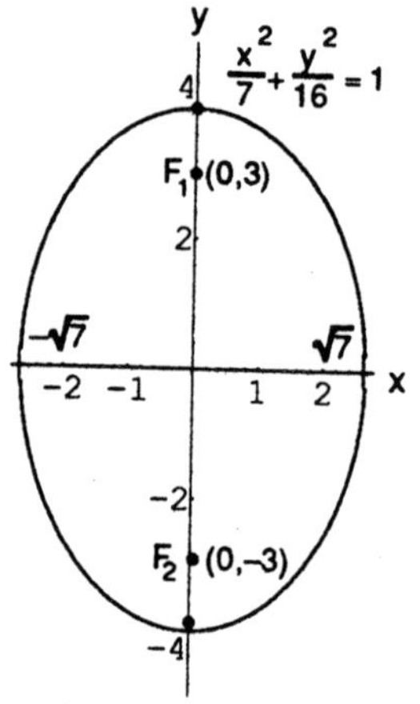

Graph 10.P.5

9. $x^2 = -12y \Rightarrow -\frac{x^2}{12} = y \Rightarrow 4p = 12 \Rightarrow p = 3 \Rightarrow$ focus is (0,–3), directrix is $y = 3$. Vertex is (0,0). $\therefore$ new vertex is (2,3), new focus is (2,0), new directrix is $y = 6$. The new equation is $(x-2)^2 = -12(y-3)$.

13. $\frac{y^2}{8}-\frac{x^2}{2}=1 \Rightarrow a=2\sqrt{2},\ b=\sqrt{2} \Rightarrow c=\sqrt{8+2}=\sqrt{10} \Rightarrow$ foci are $\left(0,\pm\sqrt{10}\right)$, vertices are $\left(0,\pm 2\sqrt{2}\right)$, center is (0,0). The asymptotes are $y=\pm 2x$. ∴ the new center is $\left(2,2\sqrt{2}\right)$, the new foci are $\left(2,2\sqrt{2}\pm\sqrt{10}\right)$, the new vertices are $\left(2,4\sqrt{2}\right)$, (2,0). The new asymptotes are $y-2\sqrt{2}=\pm 2(x-2)$. The new equation is $\frac{\left(y-2\sqrt{2}\right)^2}{8}-\frac{(x-2)^2}{2}=1$.

17. $y^2-2y+16x=-49 \Rightarrow y^2-2y+1=-16x-48 \Rightarrow (y-1)^2=-16(x+3)$, a parabola. The vertex is (−3,1). $4p=16 \Rightarrow p=4 \Rightarrow$ the focus is (−7,1). The directrix is x = 1.

21. $x^2+y^2-2x-2y=0 \Rightarrow x^2-2x+1+y^2-2y+1=2 \Rightarrow (x-1)^2+(y-1)^2=2$, a circle with center (1,1), radius $=\sqrt{2}$.

25. $B^2-4AC=3^2-4(1)(2)=1>0 \Rightarrow$ Hyperbola

27. $x^2-2xy+y^2=0 \Rightarrow (x-y)^2=0 \Rightarrow x-y=0$ or $y=x$, a straignt line (a degenerate parabola).

31. $B^2-4AC=\left(2\sqrt{3}\right)^2-4(1)(-1)=16 \Rightarrow$ Hyperbola. $\cot 2\alpha=\frac{A-C}{B}=\frac{1}{\sqrt{3}} \Rightarrow 2\alpha=\frac{\pi}{3} \Rightarrow \alpha=\frac{\pi}{6}$. $x=\frac{\sqrt{3}}{2}x'-\frac{1}{2}y'$, $y=\frac{1}{2}x'+\frac{\sqrt{3}}{2}y' \Rightarrow \left(\frac{\sqrt{3}}{2}x'-\frac{1}{2}y'\right)^2+2\sqrt{3}\left(\frac{\sqrt{3}}{2}x'-\frac{1}{2}y'\right)\left(\frac{1}{2}x'+\frac{\sqrt{3}}{2}y'\right)-\left(\frac{1}{2}x'+\frac{\sqrt{3}}{2}y'\right)^2=4 \Rightarrow$

$2x'^2-2y'^2=4 \Rightarrow x'^2-y'^2=2$

35. $x=\frac{t}{2},\ y=t+1 \Rightarrow 2x=t \Rightarrow y=2x+1$

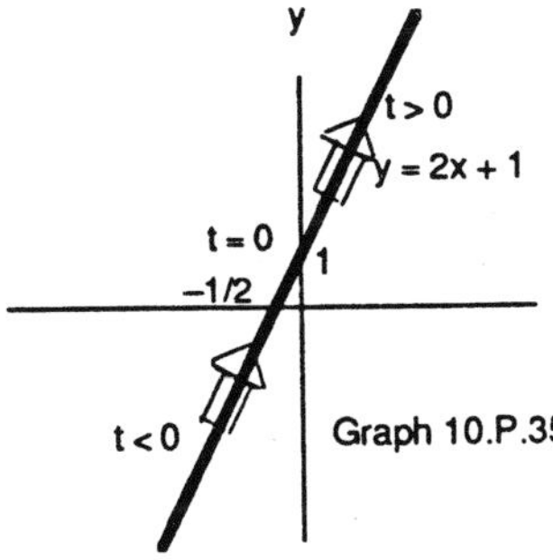

Graph 10.P.35

39. $x=-\cos t,\ y=\cos^2 t \Rightarrow y=(-x)^2=x^2$

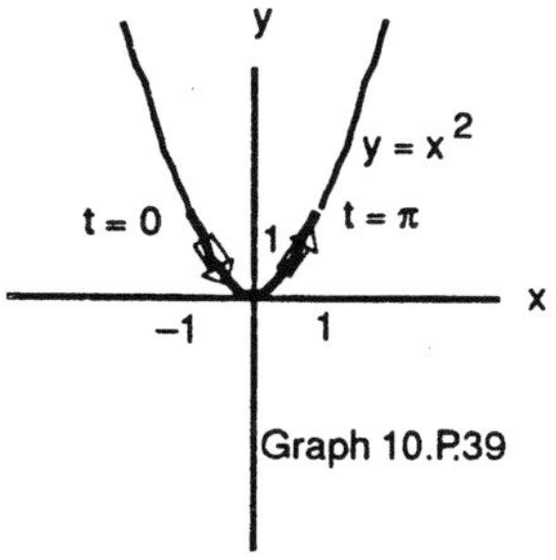

Graph 10.P39

43. $x=\frac{1}{2}\tan t,\ y=\frac{1}{2}\sec t \Rightarrow \frac{dy}{dx}=\frac{dy/dt}{dx/dt}=\frac{\frac{1}{2}\sec t\tan t}{\frac{1}{2}\sec^2 t}=\frac{\tan t}{\sec t}=\sin t \Rightarrow \frac{dy}{dx}\left(\frac{\pi}{3}\right)=\sin\frac{\pi}{3}=\frac{\sqrt{3}}{2}$. $t=\frac{\pi}{3} \Rightarrow$

$x=\frac{1}{2}\tan\frac{\pi}{3}=\frac{\sqrt{3}}{2}$ and $y=\frac{1}{2}\sec t=1 \Rightarrow y=\frac{\sqrt{3}}{2}x+\frac{1}{4}$. $\frac{d^2y}{dx^2}=\frac{dy'/dt}{dx/dt}=\frac{\cos t}{\frac{1}{2}\sec^2 t}=2\cos^3 t \Rightarrow \frac{d^2y}{dx^2}\left(\frac{\pi}{3}\right)=2\cos^3\frac{\pi}{3}=\frac{1}{4}$

47. $x=\frac{t^2}{2},\ y=2t,\ 0\le t\le\sqrt{5} \Rightarrow \frac{dx}{dt}=t,\ \frac{dy}{dt}=2 \Rightarrow \text{Area}=\int_0^{\sqrt{5}} 2\pi(2t)\sqrt{t^2+4}\,dt = \int_4^9 2\pi\,u^{1/2}\,du =$

$2\pi\left[\frac{2}{3}u^{3/2}\right]_4^9=\frac{76\pi}{3}$

Let $u=t^2+4 \Rightarrow du=2t\,dt$. $x=0 \Rightarrow u=4$, $x=\sqrt{5} \Rightarrow u=9$

51. d

55. k

59. $r = \sin\theta$, $r = 1 + \sin\theta \Rightarrow \sin\theta = 1 + \sin\theta \Rightarrow \varnothing$. There are no points of intersection found by solving the system. The point of intersection (0,0) is found by graphing.

63. $r = 1 + \sin\theta$ and $r = -1 + \sin\theta$ intersect at all points of $r = 1 + \sin\theta$. This can be seen by graphing them.

67. $r^2 = \cos 2\theta \Rightarrow r = 0$ when $\cos 2\theta = 0 \Rightarrow 2\theta = \frac{\pi}{2}, \frac{3\pi}{2} \Rightarrow \theta = \frac{\pi}{4}, \frac{3\pi}{4}$. $\theta_1 = \frac{\pi}{4} \Rightarrow m_1 = \tan\frac{\pi}{4} = 1 \Rightarrow$ $y = x$ is one tangent line. $\theta_2 = \frac{3\pi}{4} \Rightarrow m_2 = \tan\frac{3\pi}{4} = -1 \Rightarrow y = -x$ is other tangent line.

71. $r\cos\left(\theta + \frac{\pi}{3}\right) = 2\sqrt{3} \Rightarrow r\left(\cos\theta\cos\frac{\pi}{3} - \sin\theta\sin\frac{\pi}{3}\right) = 2\sqrt{3}$
$\Rightarrow \frac{1}{2} r\cos\theta - \frac{\sqrt{3}}{2} r\sin\theta = 2\sqrt{3} \Rightarrow r\cos\theta - \sqrt{3}\, r\sin\theta = 4\sqrt{3} \Rightarrow$
$x - \sqrt{3}\, y = 4\sqrt{3}$

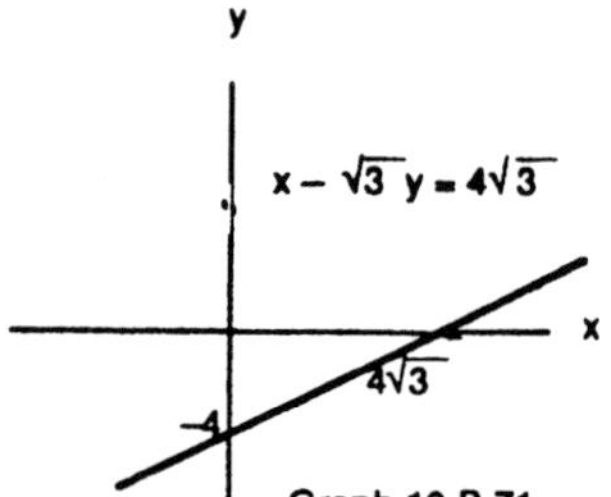

Graph 10.P.71

75. $r = -\frac{3}{2}\csc\theta \Rightarrow r\sin\theta = -\frac{3}{2} \Rightarrow y = -\frac{3}{2}$

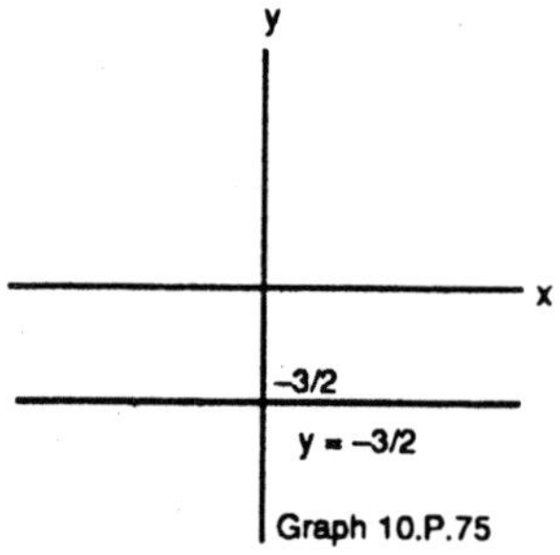

Graph 10.P.75

79. $x^2 + y^2 - 3x = 0 \Rightarrow \left(x - \frac{3}{2}\right)^2 + y^2 = \frac{9}{4} \Rightarrow C = \left(\frac{3}{2}, 0\right), a = \frac{3}{2} \Rightarrow$
$r = 3\cos\theta$ is the polar equation.

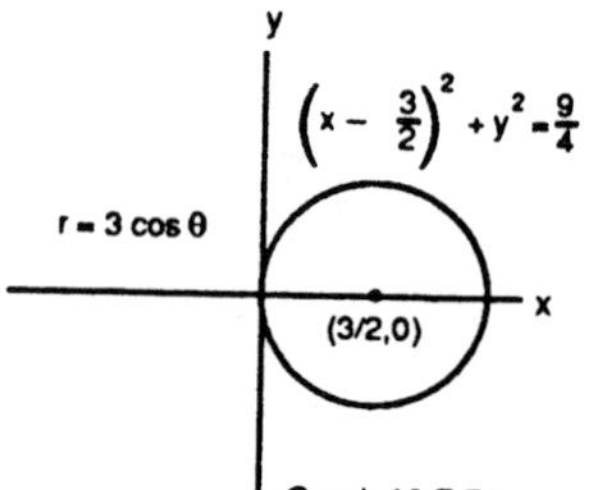

Graph 10.P.79

83. $r = 2\sqrt{2}\cos\theta \Rightarrow C = (\sqrt{2}, 0), a = \sqrt{2} \Rightarrow \left(x - \sqrt{2}\right)^2 + y^2 = 2$ is the Cartesian equation.

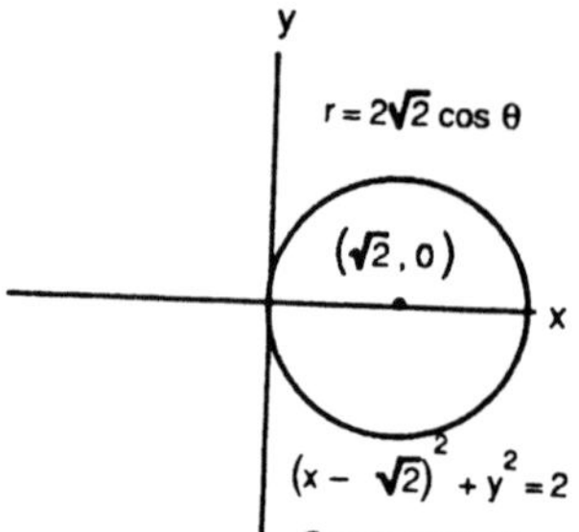

Graph 10.P.83

87. $r = \dfrac{6}{1 - 2\cos\theta} \Rightarrow e = 2 \Rightarrow$ Hyperbola

$ke = 6 \Rightarrow 2k = 6 \Rightarrow k = 3 \Rightarrow$ Vertices are $(2,\pi)$ and $(6,\pi)$.

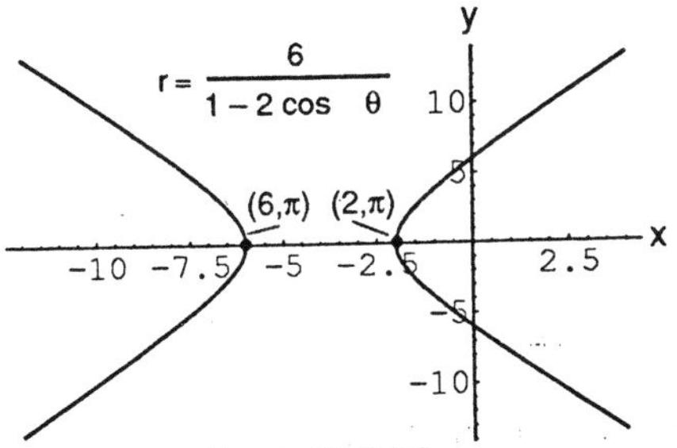

Graph 10.P.87

91. $e = \frac{1}{2}$, $r\sin\theta = 2 \Rightarrow y = 2$ is directrix $\Rightarrow k = 2$. The conic is an ellipse. $r = \dfrac{ke}{1 + e\sin\theta} \Rightarrow r = \dfrac{2\left(\frac{1}{2}\right)}{1 + \frac{1}{2}\sin\theta}$

$\Rightarrow r = \dfrac{2}{2 + \sin\theta}$

95. $r = 1 + \cos 2\theta$, $r = 1 \Rightarrow 1 = 1 + \cos 2\theta \Rightarrow 0 = \cos 2\theta \Rightarrow 2\theta = \frac{\pi}{2} \Rightarrow \theta = \frac{\pi}{4}$.

$\therefore A = 4\displaystyle\int_0^{\pi/4} \frac{1}{2}\left((1 + \cos 2\theta)^2 - 1^2\right) d\theta = 2\int_0 (1 + 2\cos 2\theta + \cos^2 2\theta - 1)\, d\theta =$

$2\displaystyle\int_0^{\pi/4} \left(2\cos 2\theta + \frac{1}{2} + \frac{\cos 4\theta}{2}\right) d\theta = 2\left[\sin 2\theta + \frac{1}{2}\theta + \frac{\sin 4\theta}{8}\right]_0^{\pi/4} = 2 + \frac{\pi}{4}$

99. $r = 8\sin^3\left(\frac{\theta}{3}\right)$, $0 \le \theta \le \frac{\pi}{4} \Rightarrow \frac{dr}{d\theta} = 8\sin^2\left(\frac{\theta}{3}\right)\cos\left(\frac{\theta}{3}\right)$. $r^2 + \left(\frac{dr}{d\theta}\right)^2 = \left(8\sin^3\left(\frac{\theta}{3}\right)\right)^2 + \left(8\sin^2\left(\frac{\theta}{3}\right)\cos\left(\frac{\theta}{3}\right)\right)^2 =$

$64\sin^4\left(\frac{\theta}{3}\right)$. $\therefore L = \displaystyle\int_0^{\pi/4} \sqrt{64\sin^4(\theta/3)}\, d\theta = \int_0^{\pi/4} 8\sin^2\left(\frac{\theta}{3}\right) d\theta = \int_0^{\pi/4} 8\left(\frac{1 - \cos\left(\frac{2\theta}{3}\right)}{2}\right) d\theta =$

$\displaystyle\int_0^{\pi/4} \left(4 - 4\cos\left(\frac{2\theta}{3}\right)\right) d\theta = \left[4\theta - 6\sin\left(\frac{2\theta}{3}\right)\right]_0^{\pi/4} = 4\left(\frac{\pi}{4}\right) - 6\sin\left(\frac{2(\pi/4)}{3}\right) - 0 = \pi - 6\sin\frac{\pi}{6} = \pi - 3.$

103. **Each portion of the wave front will reflect to the other focus, and, since the wave front travels at a constant speed as it expands, the different portions of the wave will arrive at the second focus simultaneously.**

107. a) $r = \dfrac{k}{1 + e\cos\theta} \Rightarrow r + er\cos\theta = k \Rightarrow \sqrt{x^2 + y^2} + ex = k \Rightarrow \sqrt{x^2 + y^2} = k - ex \Rightarrow x^2 + y^2 = k^2 - 2kex +$
$e^2x^2 \Rightarrow x^2 - e^2x^2 + y^2 + 2kex - k^2 = 0 \Rightarrow (1 - e^2)x^2 + y^2 + 2kex - k^2 = 0$.

b) $e = 0 \Rightarrow x^2 + y^2 - k^2 = 0 \Rightarrow$ Circle. $0 < e < 1 \Rightarrow 0 < e^2 < 1 \Rightarrow 0 < 1 - e^2 \Rightarrow x^2$ and y^2 have positive, unequal coefficients $\Rightarrow$ Ellipse.

$e = 1 \Rightarrow y^2 + 2kex - k^2 = 0 \Rightarrow$ Parabola

$e > 1 \Rightarrow e^2 > 1 \Rightarrow 1 - e^2 < 0 \Rightarrow$ the coefficient of x^2 is negative, the coefficient of y^2 is positive $\Rightarrow$ Hyperbola.

CHAPTER 11

VECTORS AND ANALYTIC GEOMETRY IN SPACE

11.1 VECTORS IN THE PLANE

1.

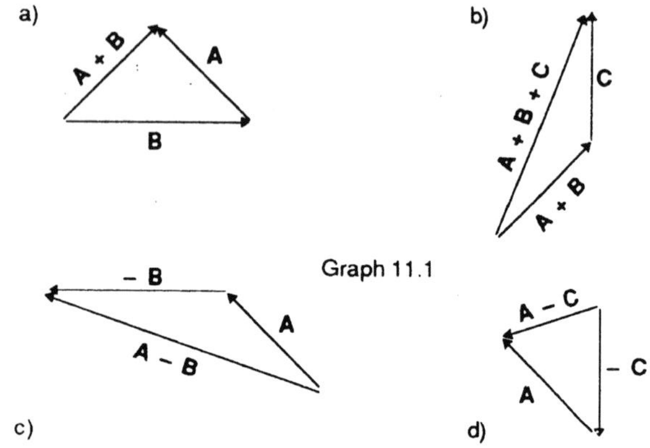

Graph 11.1

5. a) $\mathbf{w} = \mathbf{u} + \mathbf{v}$ b) $\mathbf{v} = \mathbf{w} + (-\mathbf{u}) = \mathbf{w} - \mathbf{u}$

9. $(2\mathbf{i} - 7\mathbf{j}) + (\mathbf{i} + 6\mathbf{j}) = 3\mathbf{i} - \mathbf{j}$

13. $2\Big((\ln 2)\mathbf{i} + \mathbf{j}\Big) - \Big((\ln 8)\mathbf{i} + \pi\mathbf{j}\Big) = -\ln 2\,\mathbf{i} + (2 - \pi)\mathbf{j}$

17. $\overrightarrow{AO} = -2\mathbf{i} - 3\mathbf{j}$

21. $\mathbf{u} = -\frac{\sqrt{2}}{2}\mathbf{i} - \frac{\sqrt{2}}{2}\mathbf{j}$

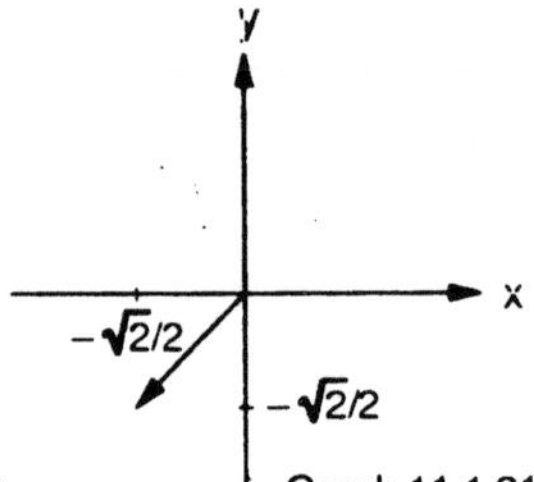

Graph 11.1.21

25. If $|\mathbf{x}|$ is the magnitude of the x–component, then $\cos 20° = \frac{|\mathbf{x}|}{|\mathbf{F}|} \Rightarrow |\mathbf{x}| = |\mathbf{F}| \cos 20° = (100) \cos 20° \approx 94$ lb $\Rightarrow$ $\mathbf{x} \approx 94\,\mathbf{i}$.

If $|\mathbf{y}|$ is the magnitude of the y–component, then $\sin 20° = \frac{|\mathbf{y}|}{|\mathbf{F}|} \Rightarrow |\mathbf{y}| = |\mathbf{F}| \sin 20° = (100) \sin 20° \approx 34.2$ lb $\Rightarrow \mathbf{y} \approx -34.2\,\mathbf{j}$ the negative sign is indicated by the diagram.

29. $\left|\sqrt{3}\,\mathbf{i} + \mathbf{j}\right| = \sqrt{\left(\sqrt{3}\right)^2 + 1^2} = 2,\ 2\left[\frac{\sqrt{3}}{2}\mathbf{i} + \frac{1}{2}\mathbf{j}\right]$

33. $\mathbf{A} = -4\mathbf{i} + 6\mathbf{j} \Rightarrow |\mathbf{A}| = \sqrt{(-4)^2 + 6^2} = \sqrt{52} = 2\sqrt{13}$ and $\mathbf{B} = 2\mathbf{i} - 3\mathbf{j} \Rightarrow |\mathbf{B}| = \sqrt{2^2 + (-3)^2} = \sqrt{13}$. Hence **A**'s direction is $\frac{\mathbf{A}}{|\mathbf{A}|} = -\frac{2}{\sqrt{13}}\mathbf{i} + \frac{3}{\sqrt{3}}\mathbf{j}$ while **B**'s is $\frac{\mathbf{B}}{|\mathbf{B}|} = \frac{2}{\sqrt{13}}\mathbf{i} - \frac{3}{\sqrt{13}}\mathbf{j} = -\frac{\mathbf{A}}{|\mathbf{A}|} \Rightarrow$ the opposite direction.

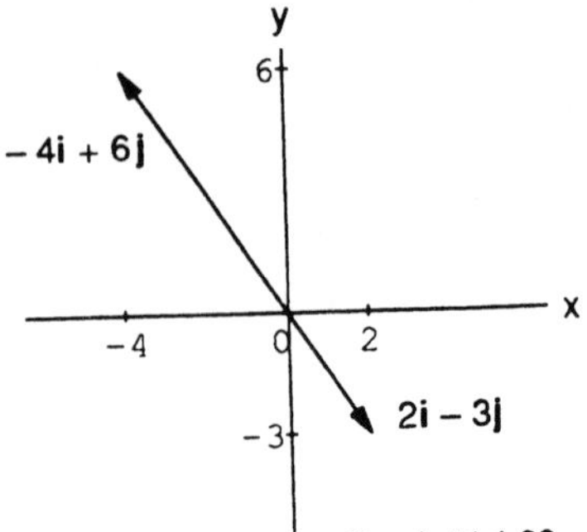

Graph 11.1.33

37. The tangent of $y = f(x)$ at $x = x_0$ has the same direction as $\mathbf{i} + f'(x_0)\,\mathbf{j}$ and the normal has the same direction as $f'(x_0)\,\mathbf{i} - \mathbf{j}$. Since $f(x) = x^2 \Rightarrow f'(x) = 2x$ at $x = 2$ the tangent has the same direction as $\mathbf{i} + 4\,\mathbf{j}$ while the normal is in the direction of $4\,\mathbf{i} - \mathbf{j}$. Therefore, the unit vectors that are tangent and normal to the curve at $x = 2$ follow. $\mathbf{u} = \frac{1}{\sqrt{17}}\mathbf{i} + \frac{4}{\sqrt{17}}\mathbf{j}$, $-\mathbf{u} = -\frac{1}{\sqrt{17}}\mathbf{i} - \frac{4}{\sqrt{17}}\mathbf{j}$ and

$\mathbf{n} = \frac{4}{\sqrt{17}}\mathbf{i} - \frac{1}{\sqrt{17}}\mathbf{j}$, $-\mathbf{n} = -\frac{4}{\sqrt{17}}\mathbf{i} + \frac{1}{\sqrt{17}}\mathbf{j}$

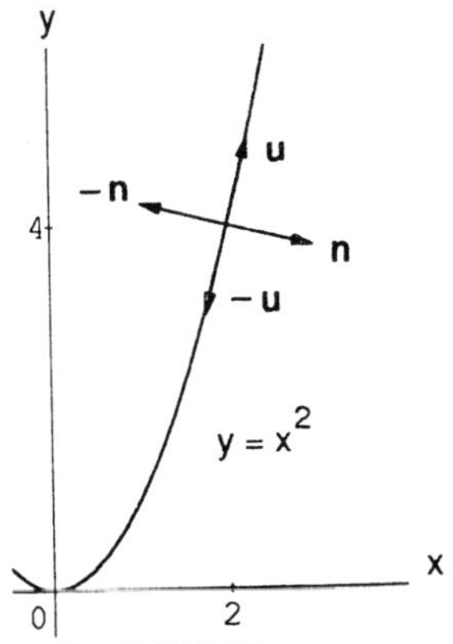

Graph 11.1.37

41. The tangent of $y = f(x)$ at $x = x_0$ has the same direction as $\mathbf{i} + f'(x_0)\,\mathbf{j}$ and the normal has the same direction as $f'(x_0)\,\mathbf{i} - \mathbf{j}$. Since $f(x) = \tan^{-1} x \Rightarrow f'(x) = \frac{1}{1 + x^2}$ at $x = 1$ the tangent has the same direction as $\mathbf{i} + \frac{1}{2}\mathbf{j}$ while the normal is in the direction of $\frac{1}{2}\mathbf{i} - \mathbf{j}$. Therefore, the unit vectors that are tangent and normal to the curve at $x = 1$ follow. $\mathbf{u} = \frac{2}{\sqrt{5}}\mathbf{i} + \frac{1}{\sqrt{5}}\mathbf{j}$, $-\mathbf{u} = -\frac{2}{\sqrt{5}}\mathbf{i} - \frac{1}{\sqrt{5}}\mathbf{j}$ and

$\mathbf{n} = -\frac{1}{\sqrt{5}}\mathbf{i} + \frac{2}{\sqrt{5}}\mathbf{j}$, $-\mathbf{n} = \frac{1}{\sqrt{5}}\mathbf{i} - \frac{2}{\sqrt{5}}\mathbf{j}$

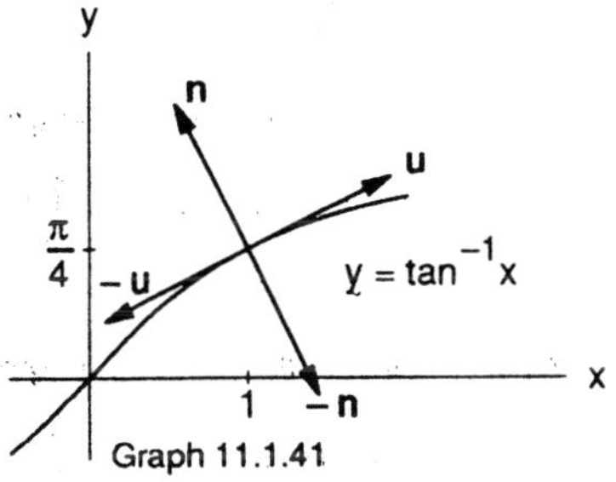

Graph 11.1.41

11.2 CARTESIAN (RECTANGULAR) COORDINATES AND VECTORS IN SPACE

1. a line through the point (2,3,0) parallel to the z–axis

5. the circle, $x^2 + y^2 = 4$ in the xy–plane

9. the circle, $y^2 + z^2 = 1$ in the yz–plane

13. a) the first quadrant of the xy–plane b) the fourth quadrant of the xy–plane

17. a) the upper hemisphere of radius 1 centered at the origin
 b) the solid upper hemisphere of radius 1 centered at the origin

21. a) $z = 1$ b) $x = 3$ c) $y = -1$

25. a) $y = 3, z = -1$ b) $x = 1, z = -1$ c) $x = 1, y = 3$

29. $0 \le z \le 1$

33. a) $(x-1)^2 + (y-1)^2 + (z-1)^2 < 1$ b) $(x-1)^2 + (y-1)^2 + (z-1)^2 > 1$

37. length $= |\mathbf{i} + 4\mathbf{j} - 8\mathbf{k}| = \sqrt{1 + 16 + 64} = 9$, the direction is $\frac{1}{9}\mathbf{i} + \frac{4}{9}\mathbf{j} - \frac{8}{9}\mathbf{k} \Rightarrow \mathbf{i} + 4\mathbf{j} - 8\mathbf{k} = 9\left[\frac{1}{9}\mathbf{i} + \frac{4}{9}\mathbf{j} - \frac{8}{9}\mathbf{k}\right]$

41. length $= \left|\frac{3}{5}\mathbf{i} + \frac{4}{5}\mathbf{k}\right| = \sqrt{\frac{9}{25} + \frac{16}{25}} = 1$, the direction is $\frac{3}{5}\mathbf{i} + \frac{4}{5}\mathbf{k} \Rightarrow \frac{3}{5}\mathbf{i} + \frac{4}{5}\mathbf{k} = 1\left[\frac{3}{5}\mathbf{i} + \frac{4}{5}\mathbf{k}\right]$

45. the distance = the length = $\left|\overrightarrow{P_1P_2}\right| = |2i + 2j - k| = \sqrt{2^2 + 2^2 + (-1)^2} = 3$, $2i + 2j - k = 3\left[\frac{2}{3}i + \frac{2}{3}j - \frac{1}{3}k\right] \Rightarrow$ the direction is $\frac{2}{3}i + \frac{2}{3}j - \frac{1}{3}k$, the midpoint is (2,2,1/2)

49. the distance = the length = $\left|\overrightarrow{P_1P_2}\right| = |2i - 2j - 2k| = \sqrt{3 \cdot 2^2} = 2\sqrt{3}$, $2i - 2j - 2k = 2\sqrt{3}\left[\frac{1}{\sqrt{3}}i - \frac{1}{\sqrt{3}}j - \frac{1}{\sqrt{3}}k\right] \Rightarrow$ the direction is $\frac{1}{\sqrt{3}}i - \frac{1}{\sqrt{3}}j - \frac{1}{\sqrt{3}}k$, the midpoint is (1,– 1,– 1)

53. $\frac{A}{|A|} = \frac{1}{13}A = \frac{1}{13}[12i - 5k] = \frac{12}{13}i - \frac{5}{13}k$, where $|A| = \sqrt{12^2 + 5^2} = \sqrt{169} = 13$; the desired vector is $\frac{7}{13}(12i - 5k)$

57. center (– 2,0,2), radius $2\sqrt{2}$

61. $(x - 1)^2 + (y - 2)^2 + (z - 3)^2 = 14$

65. $x^2 + y^2 + z^2 + 4x - 4z = 0 \Rightarrow \left(x^2 + 4x + 4\right) + \left(y^2\right) + \left(z^2 - 4z + 4\right) = 4 + 4 \Rightarrow (x + 2)^2 + y^2 + (z - 2)^2 = \left(\sqrt{8}\right)^2 \Rightarrow$ the center is at (– 2,0,2) and the radius is $\sqrt{8}$

69. a) the distance between (x,y,z) and (x,0,0) is $\sqrt{y^2 + z^2}$

b) the distance between (x,y,z) and (0,y,0) is $\sqrt{x^2 + z^2}$

c) the distance between (x,y,z) and (0,0,z) is $\sqrt{x^2 + y^2}$

11.3 DOT PRODUCTS

	A·B	|A|	|B|	cos θ	|B| cos θ	$\text{Proj}_A B$
1.	– 25	5	5	– 1	– 5	$-2i + 4j - \sqrt{5}k$
5.	0	$\sqrt{53}$	1	0	0	**0**
9.	10	$\sqrt{13}$	$\sqrt{26}$	$\frac{10}{13\sqrt{2}}$	$\frac{10}{\sqrt{13}}$	$\frac{10}{13}[3i + 2j]$

13. $B = \left(\frac{A \cdot B}{A \cdot A}A\right) + \left(B - \frac{A \cdot B}{A \cdot A}A\right) = \frac{3}{2}[i + j] + \left[(3j + 4k) - \frac{3}{2}(i + j)\right] = \left[\frac{3}{2}i + \frac{3}{2}j\right] + \left[-\frac{3}{2}i + \frac{3}{2}j + 4k\right]$, where $A \cdot B = 3$ and $A \cdot A = 2$

17. $(i + 2j) \cdot \left((x - 2)i + (y - 1)j\right) = 0 \Rightarrow x + 2y = 4$

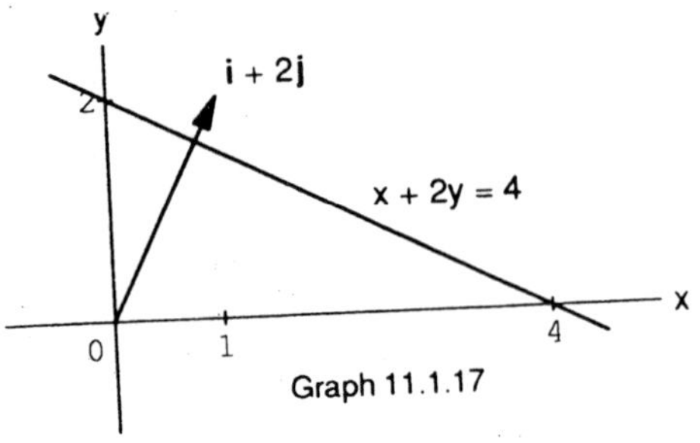

Graph 11.1.17

21. distance = $\left|\text{proj}_N \overrightarrow{PS}\right| = \left|\frac{N \cdot \overrightarrow{PS}}{|N|}\right| = \left|\frac{(i + 3j) \cdot (2i + 6j)}{\sqrt{1^2 + 3^2}}\right| = \left|\frac{20}{\sqrt{10}}\right| = 2\sqrt{10}$, where S(2,8), P(0,2) and $N = i + 3j$

25. The distance between the parallel lines is $\text{proj}_{\mathbf{N}} \overrightarrow{P_1P_2}$ where $\mathbf{N}$ is normal to the given lines and each line contains only one of the points P_1 or P_2. Let P_1 be located at (1,2), P_2 at (1,1) and $\mathbf{N} = 3\mathbf{i} + 2\mathbf{j}$. The distance is

$$\left|\text{proj}_{\mathbf{N}} \overrightarrow{P_1P_2}\right| = \left|\frac{\overrightarrow{P_1P_2} \cdot \mathbf{N}}{\mathbf{N} \cdot \mathbf{N}} \mathbf{N}\right| = \left|\frac{(-\mathbf{j}) \cdot (3\mathbf{i} + 2\mathbf{j})}{(3\mathbf{i} + 2\mathbf{j}) \cdot (3\mathbf{i} + 2\mathbf{j})} (3\mathbf{i} + 2\mathbf{j})\right| = \left|\frac{-2}{9+4} (3\mathbf{i} + 2\mathbf{j})\right| = \frac{2}{13}\sqrt{9+4} = \frac{2}{\sqrt{13}}.$$

29. a) Since $|\cos\theta| \le 1$, we have $|\mathbf{u} \cdot \mathbf{v}| = |\mathbf{u}|\,|\mathbf{v}|\,|\cos\theta| \le |\mathbf{u}|\,|\mathbf{v}|\,(1) = |\mathbf{u}|\,|\mathbf{v}|$.

b) We have equality precisely when $|\cos\theta| = 1$ or when one or both of $\mathbf{u}$ and $\mathbf{v}$ is $\mathbf{0}$. In the case of nonzero vectors, we have equality when $\theta = 0$ or π, i.e., when the vectors are parallel.

33. The sum of two vectors of equal length is *always* orthogonal to their difference, as we can see from the equation

$$(\mathbf{v}_1 + \mathbf{v}_2) \cdot (\mathbf{v}_1 - \mathbf{v}_2) = \mathbf{v}_1 \cdot \mathbf{v}_1 + \mathbf{v}_2 \cdot \mathbf{v}_1 - \mathbf{v}_1 \cdot \mathbf{v}_2 - \mathbf{v}_2 \cdot \mathbf{v}_2 = |\mathbf{v}_1|^2 - |\mathbf{v}_2|^2$$

37. Clearly the diagonals of a rectangle are equal in length. What is not as obvious is the statement that equal diagonals only happen in a rectangle. We will show this by letting the opposite sides of a parallelogram be the vectors $(v_1\mathbf{i} + v_2\mathbf{j})$ and $(u_1\mathbf{i} + u_2\mathbf{j})$. The equal diagonals of the parallellogram are $d_1 = (v_1\mathbf{i} + v_2\mathbf{j}) + (u_1\mathbf{i} + u_2\mathbf{j})$ and $d_2 = (v_1\mathbf{i} + v_2\mathbf{j}) - (u_1\mathbf{i} + u_2\mathbf{j})$. Hence $|d_1| = |d_2| \Rightarrow |(v_1\mathbf{i} + v_2\mathbf{j}) + (u_1\mathbf{i} + u_2\mathbf{j})| = |(v_1\mathbf{i} + v_2\mathbf{j}) - (u_1\mathbf{i} + u_2\mathbf{j})| \Rightarrow$ $|(v_1 + u_1)\mathbf{i} + (v_2 + u_2)\mathbf{j}| = |(v_1 - u_1)\mathbf{i} + (v_2 - u_2)\mathbf{j}| \Rightarrow \sqrt{(v_1 + u_1)^2 + (v_2 + u_2)^2} =$ $\sqrt{(v_1 - u_1)^2 + (v_2 - u_2)^2} \Rightarrow v_1^2 + 2v_1u_1 + u_1^2 + v_2^2 + 2v_2u_2 + u_2^2 = v_1^2 - 2v_1u_1 + u_1^2 + v_2^2 - 2v_2u_2 + u_2^2 \Rightarrow$ $2(v_1u_1 + v_2u_2) = -2(v_1u_1 + v_2u_2) \Rightarrow v_1u_1 + v_2u_2 = 0 \Rightarrow (v_1\mathbf{i} + v_2\mathbf{j}) \cdot (u_1\mathbf{i} + u_2\mathbf{j}) = 0 \Rightarrow$ the vectors $(v_1\mathbf{i} + v_2\mathbf{j})$ and $(u_1\mathbf{i} + u_2\mathbf{j})$ are perpendicular and the parallelogram must be a rectangle.

41. $$\theta = \cos^{-1}\left(\frac{\mathbf{A} \cdot \mathbf{B}}{|\mathbf{A}|\,|\mathbf{B}|}\right) = \cos^{-1}\left(\frac{(2)(1) + (1)(2) + (0)(-1)}{\sqrt{2^2 + 1^2 + 0^2}\sqrt{1^2 + 2^2 + (-1)^2}}\right) = \cos^{-1}\left(\frac{4}{\sqrt{5}\sqrt{6}}\right) = \cos^{-1}\left(\frac{4}{\sqrt{30}}\right) \approx 0.75 \text{ rad}$$

45. $\overrightarrow{AB} = 3\mathbf{i} + \mathbf{j} - 3\mathbf{k}$, $\overrightarrow{AC} = 2\mathbf{i} - 2\mathbf{j}$, $\overrightarrow{BA} = -3\mathbf{i} - \mathbf{j} + 3\mathbf{k}$, $\overrightarrow{CA} = -2\mathbf{i} + 2\mathbf{j}$, $\overrightarrow{CB} = \mathbf{i} + 3\mathbf{j} - 3\mathbf{k}$, $\overrightarrow{BC} = -\mathbf{i} - 3\mathbf{j} + 3\mathbf{k} \Rightarrow \angle A =$ $\cos^{-1}\left(\dfrac{\overrightarrow{AB} \cdot \overrightarrow{AC}}{|\overrightarrow{AB}|\,|\overrightarrow{AC}|}\right) = \cos^{-1}\left(\dfrac{4}{\sqrt{152}}\right) \approx 1.24 \text{ rad} \approx 71.07°$, $\angle B = \cos^{-1}\left(\dfrac{\overrightarrow{BA} \cdot \overrightarrow{BC}}{|\overrightarrow{BA}|\,|\overrightarrow{BC}|}\right) =$ $\cos^{-1}\left(\dfrac{15}{19}\right) \approx 0.66 \text{ rad} \approx 37.86°$, $\angle C = \cos^{-1}\left(\dfrac{\overrightarrow{CA} \cdot \overrightarrow{CB}}{|\overrightarrow{CA}|\,|\overrightarrow{CB}|}\right) = \cos^{-1}\left(\dfrac{4}{\sqrt{152}}\right) \approx 1.24 \text{ rad} \approx 71.07°$

49. $P(0,0,0)$, $Q(1,1,1)$ and $\mathbf{F} = 5\mathbf{k} \Rightarrow \overrightarrow{PQ} = \mathbf{i} + \mathbf{j} + \mathbf{k}$, $W = \mathbf{F} \cdot \overrightarrow{PQ} = (5\mathbf{k}) \cdot (\mathbf{i} + \mathbf{j} + \mathbf{k}) = 5 \text{ N} \cdot \text{m} = 5 \text{ J}$

53. The angle between the corresponding normals is equal to the angle between the corresponding tangents.

$$\theta = \cos^{-1}\left(\frac{\mathbf{N}_1 \cdot \mathbf{N}_2}{|\mathbf{N}_1|\,|\mathbf{N}_2|}\right) = \cos^{-1}\left(\frac{1}{\sqrt{2}}\right) = 45° = \pi/4, \text{ where } \mathbf{N}_1 = 3\mathbf{i} + \mathbf{j},\ \mathbf{N}_2 = 2\mathbf{i} - \mathbf{j}.$$

57. The angle between the corresponding normals is equal to the angle between the corresponding tangents.

$$\theta = \cos^{-1}\left(\frac{\mathbf{N}_1 \cdot \mathbf{N}_2}{|\mathbf{N}_1|\,|\mathbf{N}_2|}\right) = \cos^{-1}\left(\frac{3+4}{\sqrt{9+16}\sqrt{1+1}}\right) = \cos^{-1}\left(\frac{7}{5\sqrt{2}}\right) \approx 0.14 \text{ rad, where } \mathbf{N}_1 = 3\mathbf{i} - 4\mathbf{j},\ \mathbf{N}_2 = \mathbf{i} - \mathbf{j}.$$

61. The intersection of $y = x^3$ and $x = y^2$ don't occur where $y < 0$. Therefore, we will consider $y = x^3$ and $y = \sqrt{x}$. The points of intersection for the curves $y = x^3$ and $y = \sqrt{x}$ are (0,0) and (1,1). At (0,0) the tangent line for $y = x^3$ is $y = 0$ and the tangent line for $y = \sqrt{x}$ is $x = 0$. Therefore, the angle of intersection at (0,0) is $\frac{\pi}{2}$. At (1,1) the tangent line for $y = x^3$ is $y = 3x - 2$ and the tangent line for $y = \sqrt{x}$ is $y = \frac{1}{2}x + \frac{1}{2}$. The corresponding normal vectors are: $\mathbf{N}_1 = -3\mathbf{i} + \mathbf{j}$, $\mathbf{N}_2 = -\frac{1}{2}\mathbf{i} + \mathbf{j}$. $\theta = \cos^{-1}\left(\frac{\mathbf{N}_1 \cdot \mathbf{N}_2}{|\mathbf{N}_1|\,|\mathbf{N}_2|}\right) = \cos^{-1}\frac{1}{\sqrt{2}} = \frac{\pi}{4}$; the angle is either $45° = \frac{\pi}{4}$ or $135° = \frac{3\pi}{4}$.

11.4 CROSS PRODUCTS

1. $\mathbf{A} \times \mathbf{B} = \begin{vmatrix} \mathbf{i} & \mathbf{j} & \mathbf{k} \\ 2 & -2 & -1 \\ 1 & 0 & -1 \end{vmatrix} = 3\left[\frac{2}{3}\mathbf{i} + \frac{1}{3}\mathbf{j} + \frac{2}{3}\mathbf{k}\right] \Rightarrow$ length = 3 and the direction is $\frac{2}{3}\mathbf{i} + \frac{1}{3}\mathbf{j} + \frac{2}{3}\mathbf{k}$

$\mathbf{B} \times \mathbf{A} = \begin{vmatrix} \mathbf{i} & \mathbf{j} & \mathbf{k} \\ 1 & 0 & -1 \\ 2 & -2 & -1 \end{vmatrix} = -3\left[\frac{2}{3}\mathbf{i} + \frac{1}{3}\mathbf{j} + \frac{2}{3}\mathbf{k}\right] \Rightarrow$ length = 3 and the direction is $-\frac{2}{3}\mathbf{i} - \frac{1}{3}\mathbf{j} - \frac{2}{3}\mathbf{k}$

5. $\mathbf{A} \times \mathbf{B} = \begin{vmatrix} \mathbf{i} & \mathbf{j} & \mathbf{k} \\ 2 & 0 & 0 \\ 0 & -3 & 0 \end{vmatrix} = -6\,[\mathbf{k}] \Rightarrow$ length = 6 and the direction is $-\mathbf{k}$

$\mathbf{B} \times \mathbf{A} = \begin{vmatrix} \mathbf{i} & \mathbf{j} & \mathbf{k} \\ 0 & -3 & 0 \\ 2 & 0 & 0 \end{vmatrix} = 6\,[\mathbf{k}] \Rightarrow$ length = 6 and the direction is $\mathbf{k}$

9. $\mathbf{A} \times \mathbf{B} = \begin{vmatrix} \mathbf{i} & \mathbf{j} & \mathbf{k} \\ 1 & 0 & 0 \\ 0 & 1 & 0 \end{vmatrix} = \mathbf{k}$

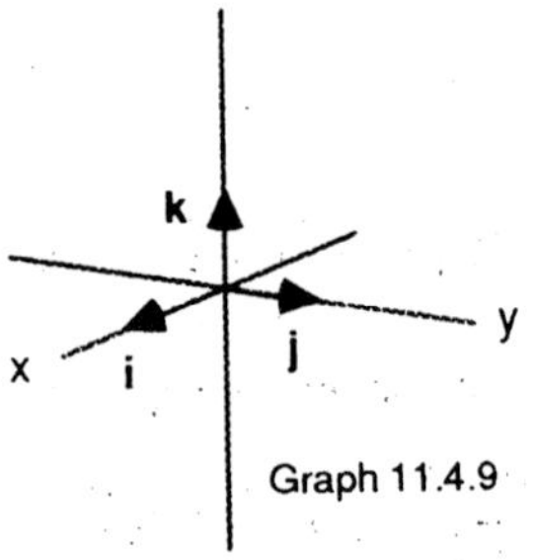

Graph 11.4.9

13. $\mathbf{A} \times \mathbf{B} = \begin{vmatrix} \mathbf{i} & \mathbf{j} & \mathbf{k} \\ 1 & 1 & 0 \\ 1 & -1 & 0 \end{vmatrix} = -2\mathbf{k}$

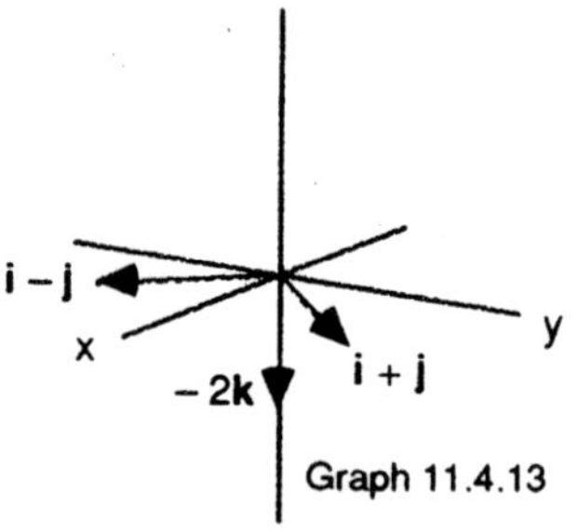

Graph 11.4.13

17. a) $\frac{\left|\overrightarrow{PQ} \times \overrightarrow{PR}\right|}{2} = \frac{\sqrt{2}}{2}$

b) $\pm\frac{\overrightarrow{PQ} \times \overrightarrow{PR}}{\left|\overrightarrow{PQ} \times \overrightarrow{PR}\right|} = \pm\left[\frac{-1}{\sqrt{2}}\mathbf{i} + \frac{1}{\sqrt{2}}\mathbf{j}\right]$

21. $\left|\overrightarrow{PQ} \times \mathbf{F}\right| = \left|\overrightarrow{PQ}\right|\,|\mathbf{F}|\sin(60°) = 10\sqrt{3}$ ft · lb

23. If $\mathbf{A} = a_1\mathbf{i} + a_2\mathbf{j} + a_3\mathbf{k}$, $\mathbf{B} = b_1\mathbf{i} + b_2\mathbf{j} + b_3\mathbf{k}$, and $\mathbf{C} = c_1\mathbf{i} + c_2\mathbf{j} + c_3\mathbf{k}$, then $\mathbf{A}\cdot(\mathbf{B}\times\mathbf{C}) = \begin{vmatrix} a_1 & a_2 & a_3 \\ b_1 & b_2 & b_3 \\ c_1 & c_2 & c_3 \end{vmatrix}$, $\mathbf{B}\cdot(\mathbf{C}\times\mathbf{A}) = \begin{vmatrix} b_1 & b_2 & b_3 \\ c_1 & c_2 & c_3 \\ a_1 & a_2 & a_3 \end{vmatrix}$ and $\mathbf{C}\cdot(\mathbf{A}\times\mathbf{B}) = \begin{vmatrix} c_1 & c_2 & c_3 \\ a_1 & a_2 & a_3 \\ b_1 & b_2 & b_3 \end{vmatrix}$ which all have the same value, since the interchanging of two pair of rows in a determinant does not change its value. The volume is $|(\mathbf{A}\times\mathbf{B})\cdot\mathbf{C}| = \begin{vmatrix} 2 & 0 & 0 \\ 0 & 2 & 0 \\ 0 & 0 & 2 \end{vmatrix} = 8$.

27. $\frac{1}{2}\left|\overrightarrow{AB}\times\overrightarrow{AC}\right| = \frac{\sqrt{4+16+16}}{2} = 3$ see example 4 section 11.4

31. No, **B** need not equal **C**. For example, $\mathbf{i}+\mathbf{j} \neq -\mathbf{i}+\mathbf{j}$, but
$\mathbf{i}\times(\mathbf{i}+\mathbf{j}) = \mathbf{i}\times\mathbf{i} + \mathbf{i}\times\mathbf{j} = \mathbf{0} + \mathbf{k} = \mathbf{k}$ and $\mathbf{i}\times(-\mathbf{i}+\mathbf{j}) = -\mathbf{i}\times\mathbf{i} + \mathbf{i}\times\mathbf{j} = \mathbf{0} + \mathbf{k} = \mathbf{k}$.

35. Let $\overrightarrow{AB} = (2+1)\mathbf{i} + (0-2)\mathbf{j} = 3\mathbf{i} - 2\mathbf{j}$; $\overrightarrow{AD} = (4+1)\mathbf{i} + (3-2)\mathbf{j} = 5\mathbf{i} + \mathbf{j}$, $\overrightarrow{AB}\times\overrightarrow{AD} = (3\mathbf{i} - 2\mathbf{j})\times(5\mathbf{i} + \mathbf{j}) = 13\mathbf{k}$.
Area $= |13\mathbf{k}| = 13$.

39. Let $\overrightarrow{AB} = (1+5)\mathbf{i} + (-2-3)\mathbf{j} = 6\mathbf{i} - 5\mathbf{j}$; $\overrightarrow{AC} = (6+5)\mathbf{i} + (-2-3)\mathbf{j} = 11\mathbf{i} - 5\mathbf{j}$, $\overrightarrow{AB}\times\overrightarrow{AC} = (6\mathbf{i} - 5\mathbf{j})\times(11\mathbf{i} - 5\mathbf{j}) = 25\mathbf{k}$. Area $= \frac{1}{2}|25\mathbf{k}| = \frac{25}{2}$.

11.5 LINES AND PLANES IN SPACE

1. the direction $\mathbf{i}+\mathbf{j}+\mathbf{k}$ and $P(3,-4,-1) \Rightarrow x = 3+t$, $y = -4+t$, $z = -1+t$

5. the direction $2\mathbf{j}+\mathbf{k}$ and $P(0,0,0) \Rightarrow x = 0$, $y = 2t$, $z = t$

9. the direction $\mathbf{i}+2\mathbf{j}+2\mathbf{k}$ and $(0,-7,0) \Rightarrow x = t$, $y = -7+2t$, $z = 2t$

13. the direction $\overrightarrow{PQ} = \mathbf{i}+\mathbf{j}+3/2\mathbf{k}$ and $P(0,0,0) \Rightarrow x = t$, $y = t$, $z = 3/2\, t$
where $0 \le t \le 1$

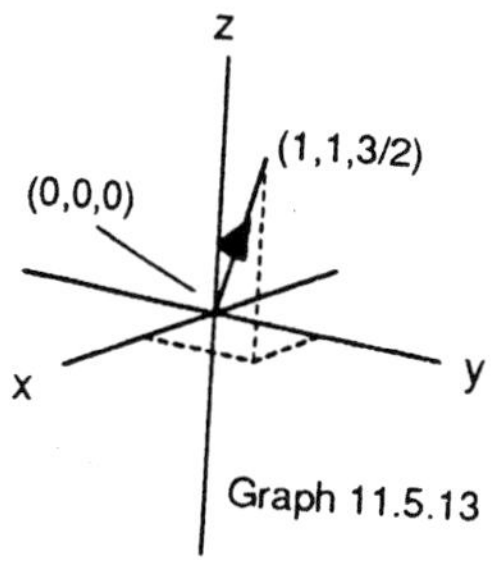

Graph 11.5.13

17. the direction $\overrightarrow{PQ} = -2\mathbf{j}$ and $P(0,1,1) \Rightarrow x = 0$, $y = 1-2t$, $z = 1$,
where $0 \le t \le 1$

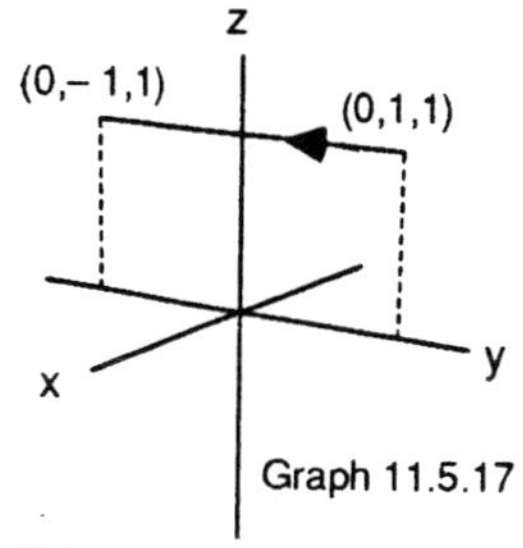

Graph 11.5.17

21. $3(x) + (-2)(y-2) + (-1)(z+1) = 0 \Rightarrow 3x - 2y - z = -3$

25. $\mathbf{N} = \mathbf{i}+3\mathbf{j}+4\mathbf{k}$, $P(2,4,5) \Rightarrow (x-2)(1) + (y-4)(3) + (z-5)(4) = 0 \Rightarrow x + 3y + 4z = 34$

29. $x = 2+2t$, $y = -4-t$, $z = 7+3t$; $x = -2-t$, $y = -2+(1/2)t$, $z = 1-(3/2)t$

33. The distance between (2 + 2t,1 + 6t,3) and (2,1,3) is $d = \sqrt{(2t)^2 + (6t)^2}$. If $f(t) = (2t)^2 + (6t)^2$ is minimized, then d is minimized. $f'(t) = 0 \Rightarrow t = 0 \Rightarrow d = 0$. This exercise can also be done with formula (7), $d = \frac{|\overrightarrow{PS} \times \mathbf{v}|}{|\mathbf{v}|}$, on page 726 of the text.

37. S(2,– 3,4), x + 2y + 2z = 13 and P(13,0,0) is on the plane $\Rightarrow \overrightarrow{PS} = -11\mathbf{i} - 3\mathbf{j} + 4\mathbf{k}$, $\mathbf{N} = \mathbf{i} + 2\mathbf{j} + 2\mathbf{k}$;
$$d = \left|\overrightarrow{PS} \cdot \frac{\mathbf{N}}{|\mathbf{N}|}\right| = \left|\frac{-11-6+8}{\sqrt{1+4+4}}\right| = 3$$

41. S(0,– 1,0), 2x + y + 2z = 4 and P(2,0,0) is on the plane $\Rightarrow \overrightarrow{PS} = -2\mathbf{i} - \mathbf{j}$, $\mathbf{N} = 2\mathbf{i} + \mathbf{j} + 2\mathbf{k}$;
$$d = \left|\overrightarrow{PS} \cdot \frac{\mathbf{N}}{|\mathbf{N}|}\right| = \left|\frac{-4-1+0}{\sqrt{4+1+4}}\right| = \frac{5}{3}$$

45. $1(1 + 2t) + 1(1 + 5t) + 1(3t) = 2 \Rightarrow t = 0 \Rightarrow (1,1,0)$

49. $\mathbf{N}_1 = 2\mathbf{i} + 2\mathbf{j} + 2\mathbf{k}$, $\mathbf{N}_2 = 2\mathbf{i} - 2\mathbf{j} - \mathbf{k} \Rightarrow \theta = \cos^{1}\left(\frac{\mathbf{N}_1 \cdot \mathbf{N}_2}{|\mathbf{N}_1|\,|\mathbf{N}_2|}\right) = \cos^{-1}\left(\frac{4-4-2}{\sqrt{12}\sqrt{9}}\right) = \cos^{-1}\left(\frac{-1}{3\sqrt{3}}\right) \approx 1.76 \text{ rad} \approx 101.1°$

53. $\mathbf{N}_1 = \mathbf{i} + \mathbf{j} + \mathbf{k}$, $\mathbf{N}_2 = \mathbf{i} + \mathbf{j} \Rightarrow \mathbf{N}_1 \times \mathbf{N}_2 = \begin{vmatrix} \mathbf{i} & \mathbf{j} & \mathbf{k} \\ 1 & 1 & 1 \\ 1 & 1 & 0 \end{vmatrix} = -\mathbf{i} + \mathbf{j}$, the direction of the desired line; (1,1,– 1) is on both planes; the desired line is $x = 1 - t$, $y = 1 + t$, $z = -1$

57. $x = 0 \Rightarrow t = -\frac{1}{2}, y = -\frac{1}{2}, z = -\frac{3}{2} \Rightarrow \left(0, -\frac{1}{2}, -\frac{3}{2}\right)$; $y = 0 \Rightarrow t = -1, x = -1, z = -3 \Rightarrow (-1,0,-3)$;
$z = 0 \Rightarrow t = 0 \Rightarrow x = 1, y = -1 \Rightarrow (1,-1,0)$

61. The cross product of $\mathbf{i} + \mathbf{j} - \mathbf{k}$ and $-4\mathbf{i} + 2\mathbf{j} - 2\mathbf{k}$ has the same direction as the normal of the plane.
$\mathbf{N} = \begin{vmatrix} \mathbf{i} & \mathbf{j} & \mathbf{k} \\ 1 & 1 & -1 \\ -4 & 2 & -2 \end{vmatrix} = 6\mathbf{j} + 6\mathbf{k}$. Select a point in either line, such as (– 1,2,1). If the point (x,y,z) is on the plane, then the vector $(x + 1)\mathbf{i} + (y - 2)\mathbf{j} + (z - 1)\mathbf{k}$ is also on the plane. Therefore, the desired plane is $\left((x + 1)\mathbf{i} + (y - 2)\mathbf{j} + (z - 1)\mathbf{k}\right) \cdot \mathbf{N} = 0 \Rightarrow 0(x + 1) + 6(y - 2) + 6(z - 1) = 0 \Rightarrow 6y - 12 + 6z - 6 = 0 \Rightarrow$
$6y + 6z + 18 \Rightarrow y + z = 3$.

11.6 SURFACES IN SPACE

1. d, ellipsoid of revolution

5. l, hyperbolic paraboloid

9. k, hyperbolic paraboloid

13. $x^2 + y^2 = 4$

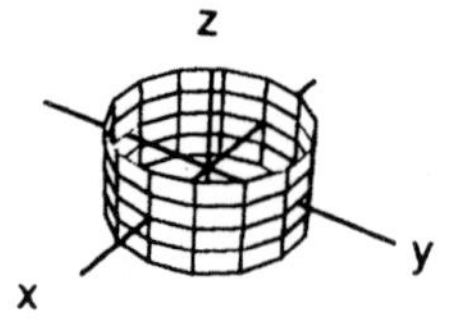

Graph 11.6.13

17. $x^2 + 4z^2 = 16$

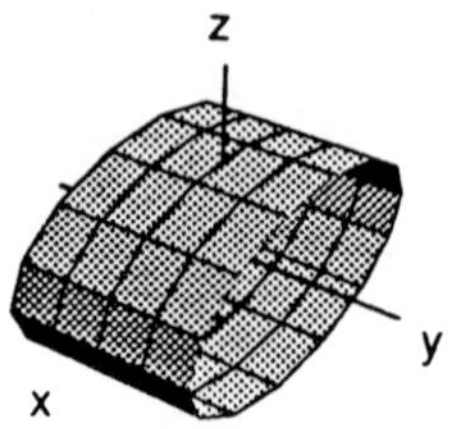

Graph 11.6.17

21. $9x^2 + y^2 + z^2 = 9$

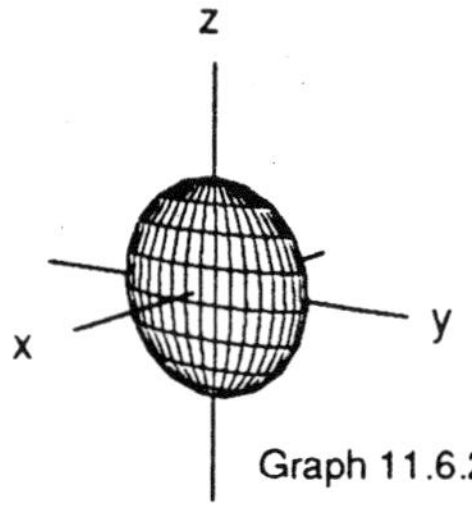

Graph 11.6.21

25. $x^2 + 4y^2 = z$

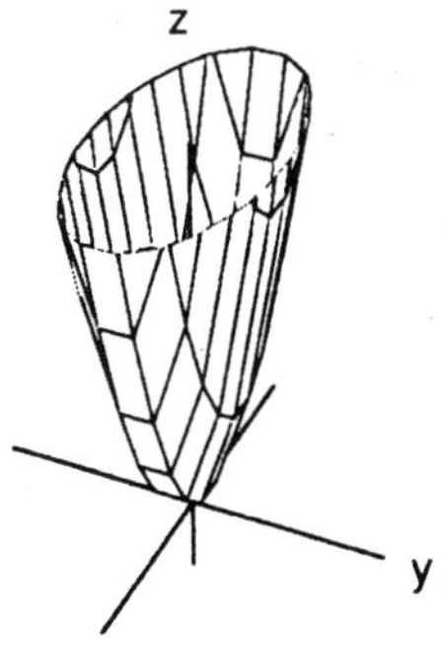

Graph 11.6.25

29. $x = 4 - 4y^2 - z^2$

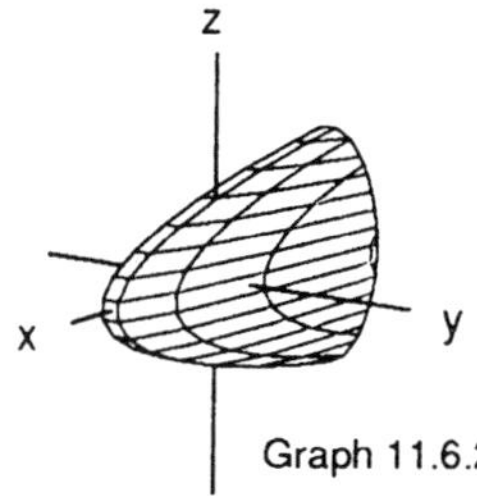

Graph 11.6.29

33. $4x^2 + 9z^2 = 9y^2$

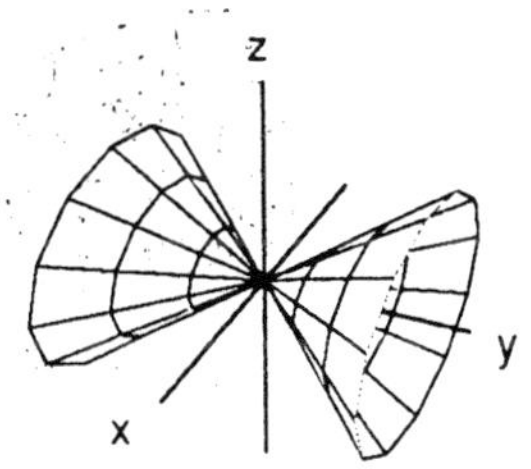

Graph 11.6.33

37. $\left(y^2/4\right) + \left(z^2/9\right) - \left(x^2/4\right) = 1$

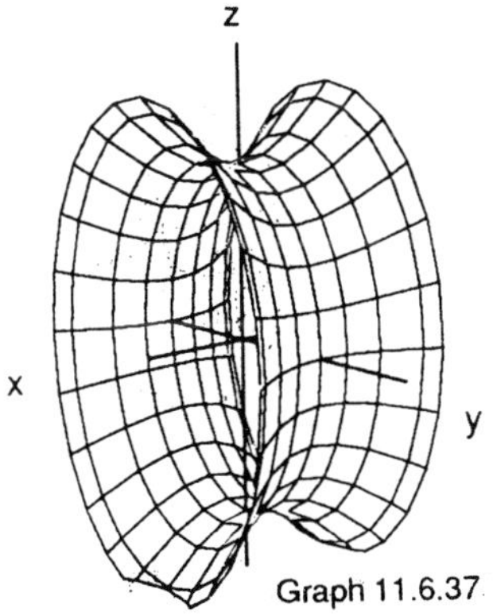

Graph 11.6.37

41. $x^2 - y^2 - \left(z^2/4\right) = 1$

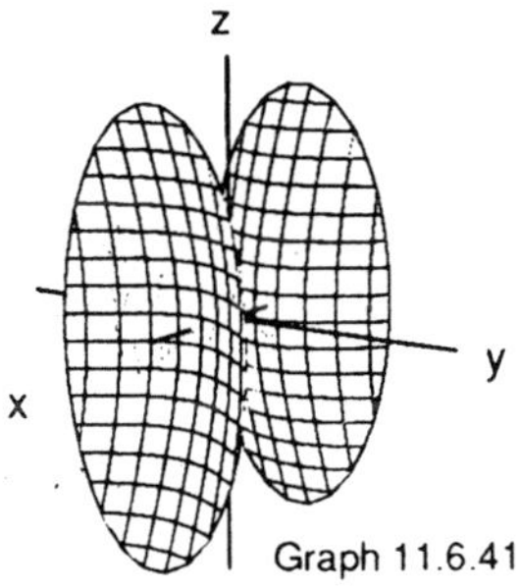

Graph 11.6.41

45. $x^2 + y^2 + z^2 = 4$

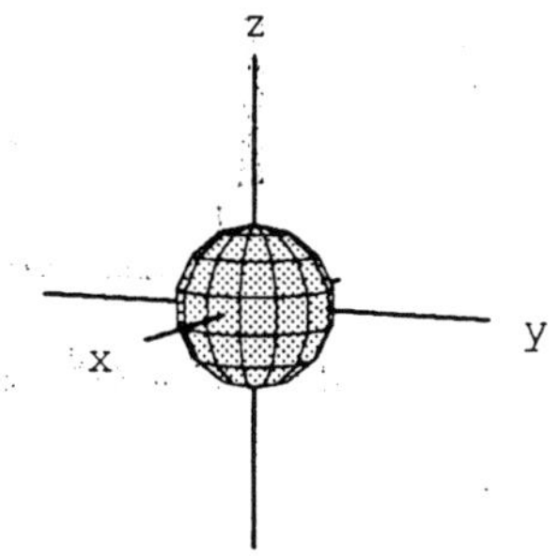

Graph 11.6.45

49. $y = -\left(x^2 + z^2\right)$

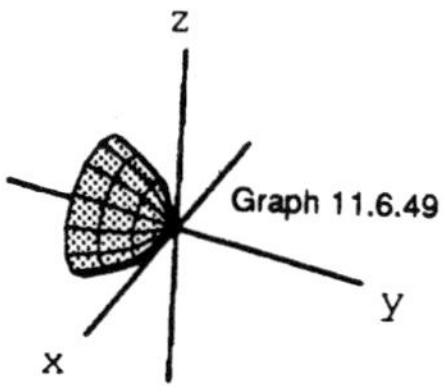

Graph 11.6.49

53. $x^2 + y^2 - z^2 = 4$

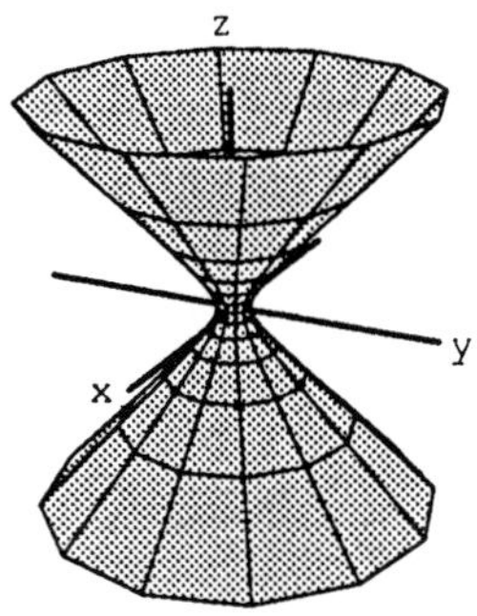

Graph 11.6.53

57. $x^2 + z^2 = 1$

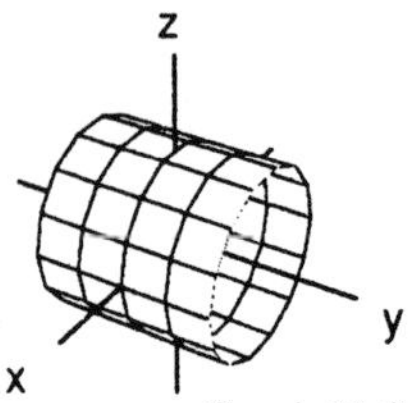

Graph 11.6.57

61. $9x^2 + 4y^2 + z^2 = 36$

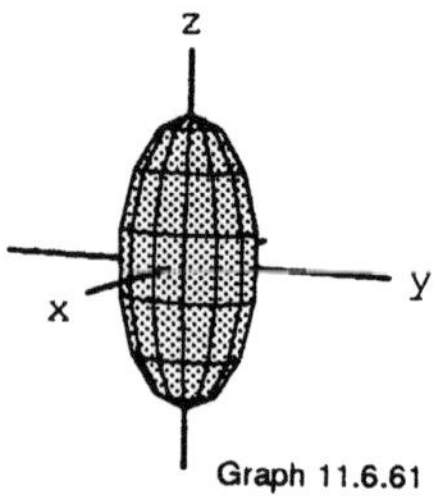

Graph 11.6.61

65. $z = -\left(x^2 + y^2\right)$

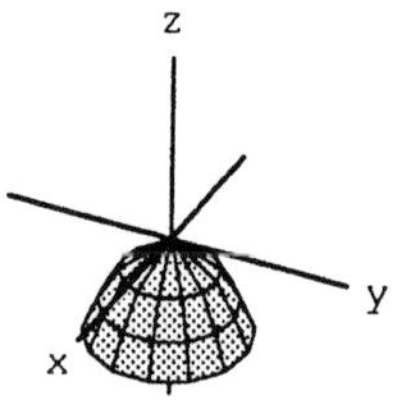

Graph 11.6.65

69. $4y^2 + z^2 - 4x^2 = 4$

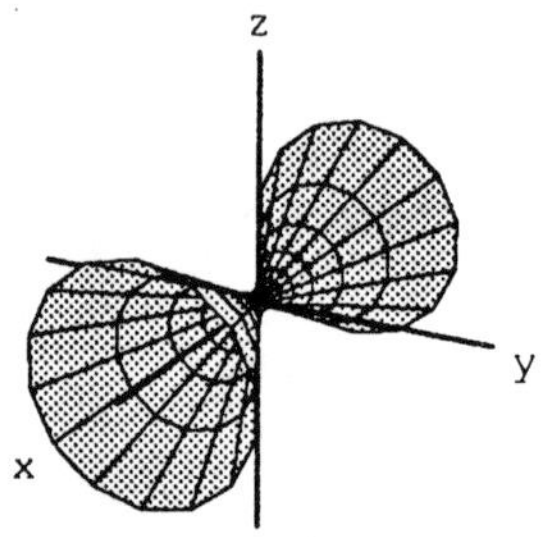

Graph 11.6.69

73. $yz = 1$

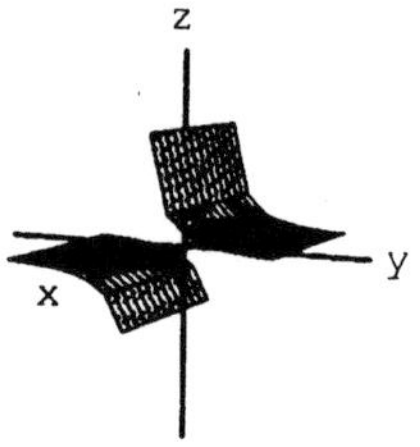

Graph 11.6.73

77. a) If $x^2 + \frac{y^2}{4} + \frac{z^2}{9} = 1$ and $z = c$, then $x^2 + \frac{y^2}{4} = \frac{9 - c^2}{9} \Rightarrow \frac{x^2}{\frac{9-c^2}{9}} + \frac{y^2}{\frac{4(9-c^2)}{9}} = 1 \Rightarrow A = ab\pi =$

$\pi\left(\frac{\sqrt{9-c^2}}{3}\right)\left(\frac{2\sqrt{9-c^2}}{3}\right) = \frac{2\pi\left(9-c^2\right)}{9}$

b) From part a) each slice has the area $\frac{2\pi\left(9-z^2\right)}{9}$ where $-3 \le z \le 3$. $\therefore$ $V = 2\int_0^3 \frac{2\pi}{9}\left(9 - z^2\right)\, dz =$

$\frac{4\pi}{9}\int_0^3 \left(9 - z^2\right) dz = \frac{4\pi}{9}\left[9z - \frac{z^3}{3}\right]_0^3 = 8\pi$

c) $\frac{x^2}{a^2} + \frac{y^2}{b^2} + \frac{z^2}{c^2} = 1 \Rightarrow \frac{x^2}{\frac{a^2\left(c^2 - z^2\right)}{c^2}} + \frac{y^2}{\frac{a^2\left(c^2 - z^2\right)}{c^2}} = 1$ and $V = 2\int_0^c \frac{\pi ab}{c^2}\left(c^2 - z^2\right) dz = \frac{4\pi abc}{3}$. If $r = a = b = c$,

then $V = \frac{4\pi r^3}{3}$ the volume of a sphere.

81. $z = x^2 + y^2$

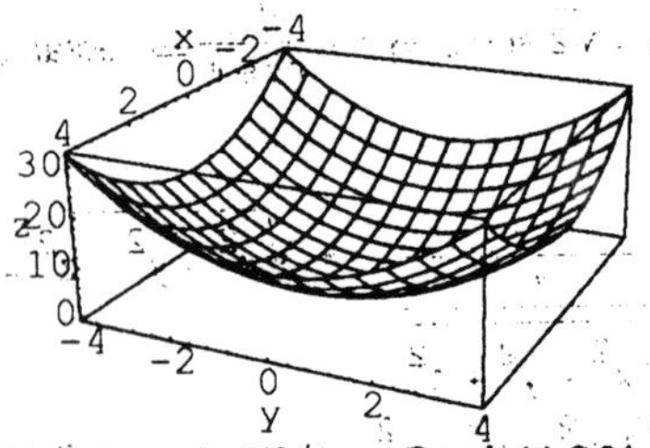

Graph 11.6.81

85. $z = \sqrt{x^2 + 2y^2 + 4}$

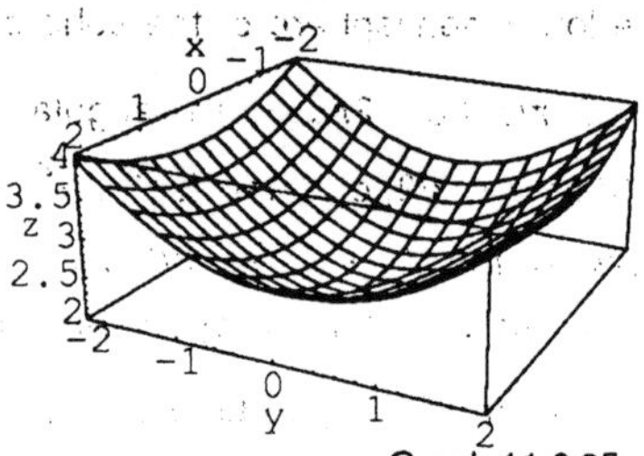

Graph 11.6.85

11.7 CYLINDRICAL AND SPHERICAL COORDINATES

	Rectangular	Cylindrical	Spherical
1.	$(0,0,0)$	$(0,0,0)$	$(0,0,0)$
5.	$(1,0,0)$	$(1,0,0)$	$(1,\pi/2,0)$
9.	$(0,-2\sqrt{2},0)$	$(2\sqrt{2},3\pi/2,0)$	$(2\sqrt{2},\pi/2,3\pi/2)$

13. $z = 0 \Rightarrow$ cylindrical, $z = 0$; spherical, $\phi = \frac{\pi}{2}$; the xy–plane

17. $\rho \sin\phi \cos\theta = 0 \Rightarrow$ rectangular, $x = 0$; cylindrical $\theta = \pi/2$; the yz–plane

21. $\rho = 5\cos\phi \Rightarrow$ rectangular, $\sqrt{x^2+y^2+z^2} = 5\cos\left(\cos^{-1}\left(\frac{z}{\sqrt{x^2+y^2+z^2}}\right)\right) \Rightarrow \sqrt{x^2+y^2+z^2} = 5\,\frac{z}{\sqrt{x^2+y^2+z^2}} \Rightarrow$
$x^2+y^2+z^2 = 5z \Rightarrow x^2+y^2+z^2-5z+\frac{25}{4} = \frac{25}{4} \Rightarrow x^2+y^2+\left(z-\frac{5}{2}\right)^2 = \left(\frac{5}{2}\right)^2$; cylindrical, $r^2+\left(z-\frac{5}{2}\right)^2 = \left(\frac{5}{2}\right)^2$,
a sphere of radius $\frac{5}{2}$ centered at $(0,0,5/2)$ (rectangular)

25. $\rho = \sqrt{2}\sec\phi \Rightarrow \rho = \frac{\sqrt{2}}{\cos\phi} \Rightarrow$ rectangular, $\sqrt{x^2+y^2+z^2} = \frac{\sqrt{2}}{\cos\left(\cos^{-1}\left(\frac{z}{\sqrt{x^2+y^2+z^2}}\right)\right)} \Rightarrow \sqrt{x^2+y^2+z^2} =$
$\frac{\sqrt{2}}{\frac{z}{\sqrt{x^2+y^2+z^2}}} \Rightarrow z\sqrt{x^2+y^2+z^2} = \sqrt{2}\sqrt{x^2+y^2+z^2} \Rightarrow \sqrt{x^2+y^2+z^2}\left(z-\sqrt{2}\right) = 0 \Rightarrow x^2+y^2+z^2 = 0$ or $z =$
$\sqrt{2}$, $x^2+y^2+z^2 \neq 0$ since $\rho = \sqrt{2}\sec\phi \neq 0$; cylindrical $z = \sqrt{2}$; the plane $z = \sqrt{2}$

29. $\rho = 3,\ \pi/3 \le \phi \le 2\pi/3 \Rightarrow$ rectangular, $\sqrt{x^2+y^2+z^2} = 3,\ 3\cos(\pi/3) \ge z \ge 3\cos(2\pi/3) \Rightarrow x^2+y^2+z^2 = 9$,
$-3/2 \le z \le 3/2$; cylindrical, $r^2+z^2 = 9,\ -3/2 \le z \le 3/2$; the portion of the sphere of radius 3 centered at the origin between the planes $z = -3/2$ and $z = 3/2$

33. $\phi = 3\pi/4,\ 0 \le \rho \le \sqrt{2} \Rightarrow$ rectangular, $\cos\phi = \cos\frac{3\pi}{4} = \frac{z}{\sqrt{x^2+y^2+z^2}} \Rightarrow -\frac{1}{\sqrt{2}} = \frac{z}{\sqrt{x^2+y^2+z^2}} \Rightarrow \sqrt{x^2+y^2+z^2} =$
$-\sqrt{2}\,z \Rightarrow x^2+y^2+z^2 = 2z^2 \Rightarrow x^2+y^2-z^2 = 0$ but $z \le 0 \Rightarrow z = -\sqrt{x^2+y^2}$, $0 \ge z \ge \sqrt{2}\cos\frac{3\pi}{4} \Rightarrow z = -\sqrt{x^2+y^2}$,
$-1 \le z \le 0$; cylindrical $x^2+y^2-z^2 = 0 \Rightarrow r^2-z^2 = 0 \Rightarrow (r+z)(r-z) = 0 \Rightarrow r = -z$ or $r = z$ but $r \ge 0$ and $z \le 0 \Rightarrow$
$r = -z$; a cone whose vertex is at the origin, the base is the circle $x^2+y^2 = 1$ in the plane $z = -1$

37. Right circular cylinder parallel to the z–axis generated by the circle
$r = -2\sin\theta$ in the $r\theta$–plane

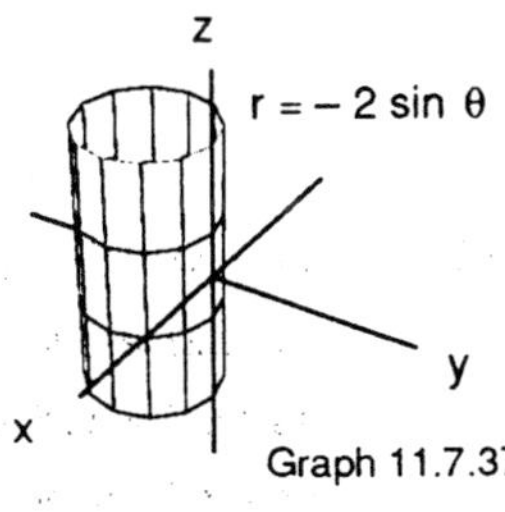

Graph 11.7.37

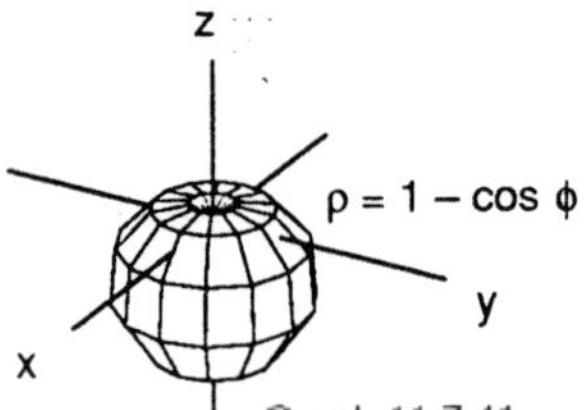

Graph 11.7.41

41. Cardioid of revolution symmetric abot the z–axis, cusp at the origin pointing down

45. a) $z = c \Rightarrow \rho\cos\phi = c \Rightarrow \rho = \frac{c}{\csc\phi} \Rightarrow \rho = c\sec\phi$

b) The xy–plane is perpendicular to the z–axis, hence $\phi = \pi/2$.

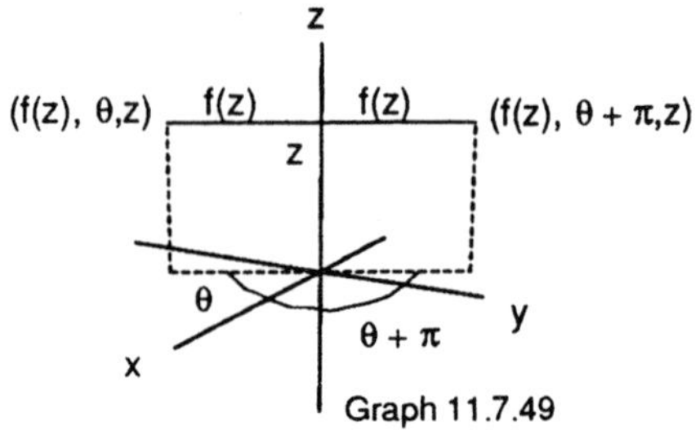

Graph 11.7.49

49. The surface's equation $r = f(z)$ tells us that the point $(r,\theta,z) = (f(z),\theta,z)$ will lie on the surface for all θ. In particular $(f(z),\theta+\pi,z)$ lies on the surface whenever $(f(z),\theta,z)$ lies on the surface, so the surface is symmetric with respect to the z–axis.

11.P PRACTICE EXERCISES

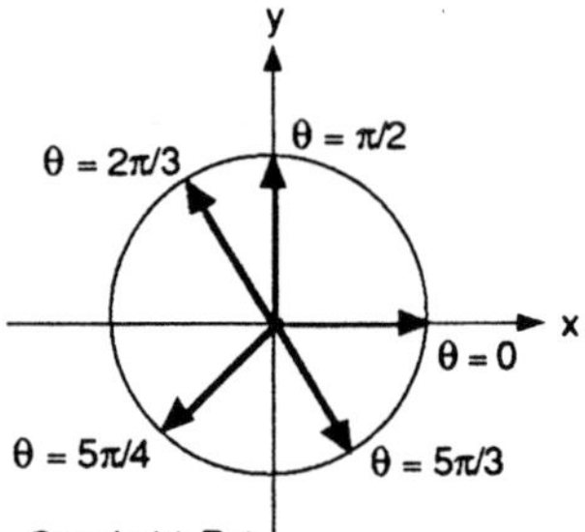

Graph 11.P.1

1. $\theta = 0 \Rightarrow \mathbf{u} = \mathbf{i}$; $\theta = \frac{\pi}{2} \Rightarrow \mathbf{u} = \mathbf{j}$; $\theta = \frac{2\pi}{3} \Rightarrow \mathbf{u} = -\frac{1}{2}\mathbf{i} + \frac{\sqrt{3}}{2}\mathbf{j}$;

$\theta = \frac{5\pi}{4} \Rightarrow \mathbf{u} = -\frac{1}{\sqrt{2}}\mathbf{i} - \frac{1}{\sqrt{2}}\mathbf{j}$; $\theta = \frac{5\pi}{3} \Rightarrow \mathbf{u} = \frac{1}{2}\mathbf{i} - \frac{\sqrt{3}}{2}\mathbf{j}$

5. length $= \left|\sqrt{2}\mathbf{i} + \sqrt{2}\mathbf{j}\right| = \sqrt{2+2} = 2$, $\sqrt{2}\mathbf{i} + \sqrt{2}\mathbf{j} = 2\left[\frac{1}{\sqrt{2}}\mathbf{i} + \frac{1}{\sqrt{2}}\mathbf{j}\right] \Rightarrow$ the direction is $\frac{1}{\sqrt{2}}\mathbf{i} + \frac{1}{\sqrt{2}}\mathbf{j}$

9. $2\frac{\mathbf{A}}{|\mathbf{A}|} = 2\frac{(4\mathbf{i} - \mathbf{j} + 4\mathbf{k})}{\sqrt{4^2 + (-1)^2 + 4^2}} = 2\frac{(4\mathbf{i} - \mathbf{j} + 4\mathbf{k})}{\sqrt{33}} = \frac{8}{\sqrt{33}}\mathbf{i} - \frac{2}{\sqrt{33}}\mathbf{j} + \frac{8}{\sqrt{33}}\mathbf{k}$

13.

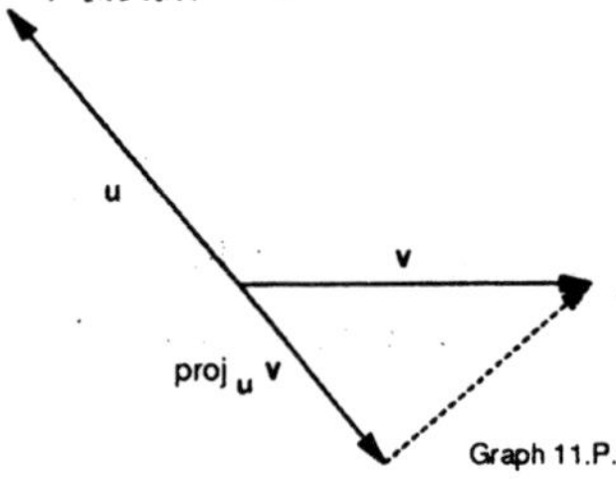

Graph 11.P.13

17. $\mathbf{B} = \left(\frac{\mathbf{A}\cdot\mathbf{B}}{\mathbf{A}\cdot\mathbf{A}}\mathbf{A}\right) + \left(\mathbf{B} - \frac{\mathbf{A}\cdot\mathbf{B}}{\mathbf{A}\cdot\mathbf{A}}\mathbf{A}\right) = \frac{4}{3}[2\mathbf{i} + \mathbf{j} - \mathbf{k}] + \left[(\mathbf{i} + \mathbf{j} - 5\mathbf{k}) - \frac{4}{3}(2\mathbf{i} + \mathbf{j} - \mathbf{k})\right] = \frac{4}{3}[2\mathbf{i} + \mathbf{j} - \mathbf{k}] - \frac{1}{3}[5\mathbf{i} + \mathbf{j} + 11\mathbf{k}]$,

where $\mathbf{A}\cdot\mathbf{B} = 8$ and $\mathbf{A}\cdot\mathbf{A} = 6$

21. Let $\mathbf{A} = a_1\mathbf{i} + a_2\mathbf{j} + a_3\mathbf{k}$ and $\mathbf{B} = b_1\mathbf{j} + b_2\mathbf{j} + b_3\mathbf{k}$. $|\mathbf{A} + \mathbf{B}|^2 + |\mathbf{A} - \mathbf{B}|^2 =$
$\left(\sqrt{(a_1 + b_1)^2 + (a_2 + b_2)^2 + (a_3 + b_3)^2}\right)^2 + \left(\sqrt{(a_1 - b_1)^2 - (a_2 - b_2)^2 - (a_3 - b_3)^2}\right)^2 = (a_1 + b_1)^2 +$
$(a_2 + b_2)^2 + (a_3 + b_3)^2 + (a_1 - b_1)^2 - (a_2 - b_2)^2 - (a_3 - b_3)^2 = a_1^2 + 2a_1b_1 + b_1^2 + a_2^2 + 2a_2b_2 + b_2^2 +$
$a_3^2 + 2a_3b_3 + b_3^2 + a_1^2 - 2a_1b_1 + b_1^2 + a_2^2 - 2a_2b_2 + b_2^2 + a_3^2 - 2a_3b_3 + b_3^2 = 2\left(a_1^2 + a_2^2 + a_3^2\right) +$
$2\left(b_1^2 + b_2^2 + b_3^2\right) = 2|\mathbf{A}|^2 + 2|\mathbf{B}|^2.$

25. The desired vector is $\mathbf{N} \times \mathbf{v}$ or $\mathbf{v} \times \mathbf{N}$. $\mathbf{N} \times \mathbf{v}$ is perpendicular to both $\mathbf{N}$ and $\mathbf{v}$ and, therefore, also parallel with the plane.

29. The desired distance $d(t) = \sqrt{(-t-2)^2 + (t-2)^2 + (t-1)^2}$ is minimized when $f(t) = (-t-2)^2 + (t-2)^2 + (t-1)^2$ is minimized. $f'(t) = 2(t+2) + 2(t-2) + 2(t-1) = 0 \Rightarrow t = \frac{1}{3}$. $\therefore$ the distance is $d(1/3) =$
$\sqrt{\left(\frac{7}{3}\right)^2 + \left(\frac{-5}{3}\right)^2 + \left(\frac{-2}{3}\right)^2} = \frac{\sqrt{78}}{3}$. This exercise can also be done with formula (7), $d = \frac{|\overrightarrow{PS} \times \mathbf{v}|}{|\mathbf{v}|}$, on page 726 of the text.

33. $S(6,0,-6)$, $P(4,0,0)$ is on $x - y = 4 \Rightarrow \overrightarrow{PS} = 2\mathbf{i} - 6\mathbf{k}$ and $\mathbf{N} = \mathbf{i} - \mathbf{j} \Rightarrow d = \left|\frac{\mathbf{N} \cdot \overrightarrow{PS}}{|\mathbf{N}|}\right| = \left|\frac{2}{\sqrt{2}}\right| = \sqrt{2}$

37. $P(1,-1,2)$, $Q(2,1,3)$ and $R(-1,2,-1) \Rightarrow \overrightarrow{PQ} = \mathbf{i} + 2\mathbf{j} + \mathbf{k}$, $\overrightarrow{PR} = -2\mathbf{i} + 3\mathbf{j} - 3\mathbf{k}$ and $\overrightarrow{PQ} \times \overrightarrow{PR} = \begin{vmatrix} \mathbf{i} & \mathbf{j} & \mathbf{k} \\ 1 & 2 & 1 \\ -2 & 3 & -3 \end{vmatrix} =$
$-9\mathbf{i} + \mathbf{j} + 7\mathbf{k}$, the normal of the plane; $-9(x-1) + 1(y+1) + 7(z-2) = 0 \Rightarrow -9x + y + 7z = 4$

41. $\mathbf{N}_1 = \mathbf{i}$ and $\mathbf{N}_2 = \mathbf{i} + \mathbf{j} + \sqrt{2}\mathbf{k} \Rightarrow$ the desired angle is $\cos^{-1}\left(\frac{\mathbf{N}_1 \cdot \mathbf{N}_2}{|\mathbf{N}_1|\,|\mathbf{N}_2|}\right) = \cos^{-1}\left(\frac{1}{2}\right) = \frac{\pi}{3}$

desired line is $x = -5 + 5t$, $y = 3 - t$, $z = -3t$.

45. a) The corresponding normals are $\mathbf{N}_1 = 3\mathbf{i} + 6\mathbf{k}$ and $\mathbf{N}_2 = 2\mathbf{i} + 2\mathbf{j} - \mathbf{k}$ and since $\mathbf{N}_1 \cdot \mathbf{N}_2 = (3)(2) + (0)(2) + (6)(-1) = 6 + 0 - 6 = 0 \Rightarrow$ the planes are orthogonal

b) The line of intersection is parallel with $\mathbf{N}_1 \times \mathbf{N}_2 = \begin{vmatrix} \mathbf{i} & \mathbf{j} & \mathbf{k} \\ 3 & 0 & 6 \\ 2 & 2 & -1 \end{vmatrix} = -12\mathbf{i} + 15\mathbf{j} + 6\mathbf{k}$. Now to find a point in the intersection, solve $\begin{cases} 3x + 6z = 1 \\ 2x + 2y - z = 3 \end{cases} \Rightarrow \begin{cases} 3x + 6z = 1 \\ 12x + 12y - 6z = 18 \end{cases} \Rightarrow 15x + 12y = 19 \Rightarrow x = 0$ and $y = \frac{19}{12} \Rightarrow (0,19/12,1/6)$ is a point on the line we seek. Therefore, the line is $x = -12t$, $y = \frac{19}{12} + 15t$ and $z = \frac{1}{6} + 6t$.

49. $W = \mathbf{F} \cdot \overrightarrow{PQ} = |\mathbf{F}|\,\left|\overrightarrow{PQ}\right| \cos\theta = (160\text{ N})(250\text{ m}) \cos\frac{\pi}{6} \approx 34641\text{ N}\cdot\text{m} = 34641\text{ J}$

53. a) true, $\sqrt{\mathbf{A} \cdot \mathbf{A}} = \sqrt{|\mathbf{A}|\,|\mathbf{A}| \cos 0} = \sqrt{|\mathbf{A}|^2} = |\mathbf{A}|$ b) not always true, $(2\mathbf{i}) \cdot (2\mathbf{i}) = 4$ while $|2\mathbf{i}| = \sqrt{2^2} = 2$

c) true, $\mathbf{A} \times \mathbf{0} = \mathbf{n}\,|\mathbf{A}|\,|\mathbf{0}| \sin\theta = \mathbf{0} = \mathbf{n}\,|\mathbf{0}|\,|\mathbf{A}| \sin\theta = \mathbf{0} \times \mathbf{A}$ d) true, $\mathbf{A} \times (-\mathbf{A}) = \mathbf{n}\,|\mathbf{A}|\,|-\mathbf{A}| \sin\pi = \mathbf{0}$

e) not always true, they may have opposite directions f) true, by the Vector Distributive Law

g) true, $\mathbf{B}$ is perpendicular to $(\mathbf{A} \times \mathbf{B})$, therefore $(\mathbf{A} \times \mathbf{B}) \cdot \mathbf{B} = 0$ h) true, a property of the Triple Scalar Product

57. $4x^2 + 4y^2 + z^2 = 4$

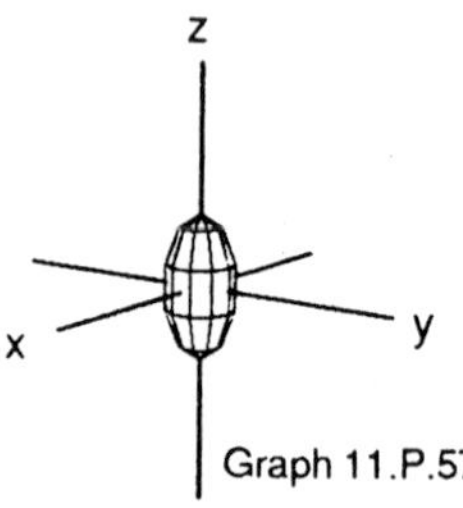

Graph 11.P.57

61. $x^2 + y^2 = z^2$

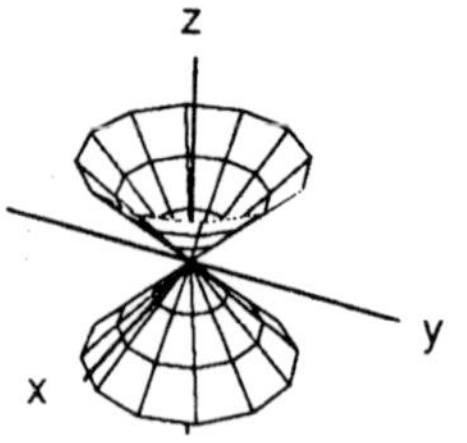

Graph 11.P.61

65. $y^2 - x^2 - z^2 = 1$

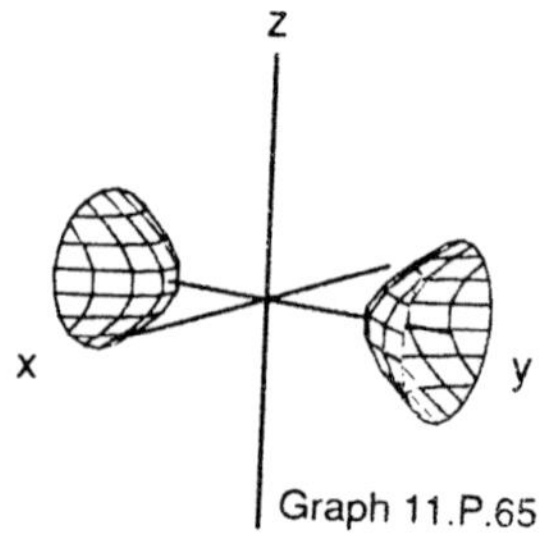

Graph 11.P.65

69. The circle centered at (0,0) with a radius of 2 in the xy–plane; the cylinder parallel with the z–axis in three dimensional space with the circle as a generating curve

73. A cardioid in the rθ–plane; a cylinder parallel with the z–axis in three dimensional space with the cardioid as a generating curve

77. The sphere with a radius of 2 centered at the origin

81. The upper hemisphere of a sphere of radius 1 centered at the origin

	Rectangular	Cylindrical	Spherical
85.	(0,1,1)	$(1,\pi/2,1)$	$(\sqrt{2},\pi/4,\pi/2)$

89. $z = 2 \Rightarrow$ cylindrical, $z = 2$; spherical, $\rho \cos \phi = 2$; a plane parallel with the xy–plane

93. $z = r^2 \Rightarrow$ rectangular, $z = x^2 + y^2$; spherical, $\rho \cos \phi = \rho^2 \sin^2\phi \Rightarrow \rho^2 \sin^2\phi - \rho \cos \phi = 0 \Rightarrow \rho \left(\rho \sin^2\phi - \cos \phi\right) = 0 \Rightarrow$ $\rho = \dfrac{\cos \phi}{\sin^2\phi}$, $0 < \phi \le \frac{\pi}{2}$, $\rho = 0$ is not needed because at $\phi = \frac{\pi}{2}$ the origin is included, a circular paraboloid symmetric to the z–axis opening upward with its vertex at the origin

97. $\rho = 4 \Rightarrow$ rectangular, $\sqrt{x^2 + y^2 + z^2} = 4 \Rightarrow x^2 + y^2 + z^2 = 16$; cylindrical, $r^2 + z^2 = 16$, a sphere of radius 4 centered at the origin

CHAPTER 12

VECTOR–VALUED FUNCTIONS

12.1 VECTOR–VALUED FUNCTIONS AND SPACE CURVES

1. $\mathbf{r} = (2\cos t)\,\mathbf{i} + (3\sin t)\,\mathbf{j} + 4t\,\mathbf{k} \Rightarrow \mathbf{v} = \frac{d\mathbf{r}}{dt} = (-2\sin t)\,\mathbf{i} + (3\cos t)\,\mathbf{j} + 4\,\mathbf{k}$

$\mathbf{a} = \frac{d^2\mathbf{r}}{dt^2} = (-2\cos t)\,\mathbf{i} - (3\sin t)\,\mathbf{j}$. Speed: $\left|\mathbf{v}\left(\frac{\pi}{2}\right)\right| = \sqrt{\left(-2\sin\frac{\pi}{2}\right)^2 + \left(3\cos\frac{\pi}{2}\right)^2 + 4^2} = 2\sqrt{5}$

Direction: $\dfrac{\mathbf{v}\left(\frac{\pi}{2}\right)}{\left|\mathbf{v}\left(\frac{\pi}{2}\right)\right|} = \left(-\frac{2}{2\sqrt{5}}\sin\frac{\pi}{2}\right)\mathbf{i} + \left(\frac{3}{2\sqrt{5}}\cos\frac{\pi}{2}\right)\mathbf{j} + \frac{4}{2\sqrt{5}}\mathbf{k} = -\frac{1}{\sqrt{5}}\mathbf{i} + \frac{2}{\sqrt{5}}\mathbf{k}$

$\mathbf{v}\left(\frac{\pi}{2}\right) = 2\sqrt{5}\left[-\frac{1}{\sqrt{5}}\,\mathbf{i} + \frac{2}{\sqrt{5}}\,\mathbf{k}\right]$

5. $\mathbf{r} = (2\ln(t+1))\,\mathbf{i} + t^2\,\mathbf{j} + \frac{t^2}{2}\,\mathbf{k} \Rightarrow \mathbf{v} = \frac{d\mathbf{r}}{dt} = \left(\frac{2}{t+1}\right)\mathbf{i} + 2t\,\mathbf{j} + t\,\mathbf{k},\ \mathbf{a} = \frac{d^2\mathbf{r}}{dt^2} = \left(\frac{-2}{(t+1)^2}\right)\mathbf{i} + 2\,\mathbf{j} + \mathbf{k}$

Speed: $|\mathbf{v}(1)| = \sqrt{\left(\frac{2}{1+1}\right)^2 + (2(1))^2 + 1^2} = \sqrt{6}$. Direction: $\dfrac{\mathbf{v}(1)}{|\mathbf{v}(1)|} = \dfrac{\left(\frac{2}{1+1}\right)\mathbf{i} + 2(1)\,\mathbf{j} + (1)\,\mathbf{k}}{\sqrt{6}}$

$= \frac{1}{\sqrt{6}}\mathbf{i} + \frac{2}{\sqrt{6}}\mathbf{j} + \frac{1}{\sqrt{6}}\mathbf{k}$. $\mathbf{v}(1) = \sqrt{6}\left[\frac{1}{\sqrt{6}}\,\mathbf{i} + \frac{2}{\sqrt{6}}\,\mathbf{j} + \frac{1}{\sqrt{6}}\,\mathbf{k}\right]$

9. $\mathbf{v} = 3\,\mathbf{i} + \sqrt{3}\,\mathbf{j} + 2t\,\mathbf{k},\ \mathbf{a} = 2\,\mathbf{k}$. $\mathbf{v}(0) = 3\,\mathbf{i} + \sqrt{3}\,\mathbf{j},\ \mathbf{a}(0) = 2\,\mathbf{k} \Rightarrow |\mathbf{v}(0)| = \sqrt{3^2 + \left(\sqrt{3}\right)^2 + 0^2} = \sqrt{12}$,
$|\mathbf{a}(0)| = \sqrt{2^2} = 2$. $\mathbf{v}(0)\cdot\mathbf{a}(0) = 0 \Rightarrow \cos\theta = \frac{0}{2\sqrt{12}} = 0 \Rightarrow \theta = \frac{\pi}{2}$

13. $\mathbf{v} = (1-\cos t)\,\mathbf{i} + (\sin t)\,\mathbf{j},\ \mathbf{a} = (\sin t)\,\mathbf{i} + (\cos t)\,\mathbf{j} \Rightarrow \mathbf{v}\cdot\mathbf{a} = \sin t(1-\cos t) + \sin t(\cos t) = \sin t$.
$\mathbf{v}\cdot\mathbf{a} = 0 \Rightarrow \sin t = 0 \Rightarrow t = 0, \pi, 2\pi$

17. $\displaystyle\int_{-\pi/4}^{\pi/4}\left((\sin t)\,\mathbf{i} + (1+\cos t)\,\mathbf{j} + (\sec^2 t)\,\mathbf{k}\right)dt = \left[-\cos t\right]_{-\pi/4}^{\pi/4}\mathbf{i} + \left[t + \sin t\right]_{-\pi/4}^{\pi/4}\mathbf{j} + \left[\tan t\right]_{-\pi/4}^{\pi/4}\mathbf{k} =$
$\left(\frac{\pi + 2\sqrt{2}}{2}\right)\mathbf{j} + 2\,\mathbf{k}$

19. $\displaystyle\int_1^4\left(\frac{1}{t}\,\mathbf{i} + \frac{1}{5-t}\,\mathbf{j} + \frac{1}{2t}\,\mathbf{k}\right)dt = \left[\ln t\right]_1^4\mathbf{i} + \left[-\ln(5-t)\right]_1^4\mathbf{j} + \left[\frac{1}{2}\ln t\right]_1^4\mathbf{k} = (\ln 4)\,\mathbf{i} + (\ln 4)\,\mathbf{j} + (\ln 2)\,\mathbf{k}$

23. $\mathbf{v} = (1 - \cos t)\,\mathbf{i} + (\sin t)\,\mathbf{j}$, $\mathbf{a} = (\sin t)\,\mathbf{i} + (\cos t)\,\mathbf{j} \Rightarrow$ For $t = \pi$, $\mathbf{v}(\pi) = 2\,\mathbf{i}$, $\mathbf{a}(\pi) = -\mathbf{j}$; For $t = \frac{3\pi}{2}$,

$\mathbf{v}\left(\frac{3\pi}{2}\right) = \mathbf{i} - \mathbf{j}$, $\mathbf{a}\left(\frac{3\pi}{2}\right) = -\mathbf{i}$

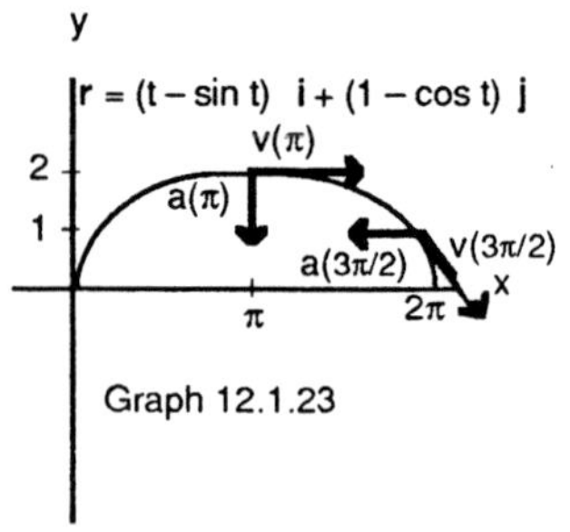

Graph 12.1.23

27. $\mathbf{r} = \int \left(\frac{3}{2}(t+1)^{1/2}\,\mathbf{i} + e^{-t}\,\mathbf{j} + \frac{1}{t+1}\,\mathbf{k}\right) dt = (t+1)^{3/2}\,\mathbf{i} - e^{-t}\,\mathbf{j} + \ln(t+1)\,\mathbf{k} + \mathbf{C}$. $\mathbf{r}(0) = \mathbf{k} \Rightarrow$

$(0+1)^{3/2}\,\mathbf{i} - e^{-0}\,\mathbf{j} + \ln(0+1)\,\mathbf{k} + \mathbf{C} = \mathbf{k} \Rightarrow \mathbf{C} = -\mathbf{i} + \mathbf{j} + \mathbf{k}$.

$\therefore\ \mathbf{r} = \left((t+1)^{3/2} - 1\right)\mathbf{i} + \left(1 - e^{-t}\right)\mathbf{j} + (1 + \ln(t+1))\,\mathbf{k}$

31. $\mathbf{v} = (1 - \cos t)\,\mathbf{i} + (\sin t)\,\mathbf{j}$, $\mathbf{a} = (\sin t)\,\mathbf{i} + (\cos t)\,\mathbf{j}$. $|\mathbf{v}|^2 = (1 - \cos t)^2 + \sin^2 t = 2 - 2\cos t$. $|\mathbf{v}|^2$ is at a max when $\cos t = -1 \Rightarrow t = \pi, 3\pi, 5\pi$, etc. At these values of t, $|\mathbf{v}|^2 = 4 \Rightarrow \max|\mathbf{v}| = \sqrt{4} = 2$. $|\mathbf{v}|^2$ is at a min when $\cos t = 1 \Rightarrow t = 0, 2\pi, 4\pi$, etc. At these values of t, $|\mathbf{v}|^2 = 0 \Rightarrow \min|\mathbf{v}| = 0$.

$|\mathbf{a}|^2 = \sin^2 t + \cos^2 t = 1$ for every $t \Rightarrow \max|\mathbf{a}| = \min|\mathbf{a}| = \sqrt{1} = 1$.

35. $\mathbf{r}(t) = (\sin t)\mathbf{i} + \left(t^2 - \cos t\right)\mathbf{j} + e^t\,\mathbf{k} \Rightarrow \mathbf{v}(t) = (\cos t)\mathbf{i} + (2t + \sin t)\mathbf{j} + e^t\,\mathbf{k}$. $t_0 = 0 \Rightarrow \mathbf{v}_0 = \mathbf{i} + \mathbf{k}$. $P_0 = (0,-1,1)$.

$\therefore$ $x = 0 + t = t$, $y = -1$, $z = 1 + t$ are the parametric equations of the tangent line.

39. Let $\mathbf{r} = \mathbf{C}$, a constant vector. Then $\mathbf{r} = c_1\,\mathbf{i} + c_2\,\mathbf{j} + c_3\,\mathbf{k}$ where c_1, c_2, c_3 are Real Numbers.

$\frac{d\mathbf{r}}{dt} = 0\,\mathbf{i} + 0\,\mathbf{j} + 0\,\mathbf{k} = 0$

43. a) Let u, $\mathbf{r}$ be continuous on [a,b]. Then $\lim_{t \to t_0} u(t)\,\mathbf{r}(t) = \lim_{t \to t_0} [u(t)\,f(t)\,\mathbf{i} + u(t)\,g(t)\,\mathbf{j} + u(t)\,h(t)\,\mathbf{k}] = u(t_0)\,f(t_0)\,\mathbf{i} + u(t_0)\,g(t_0)\,\mathbf{j} + u(t_0)\,h(t_0)\,\mathbf{k} = u(t_0)\,\mathbf{r}(t_0)$

b) Let u, $\mathbf{r}$ be differentiable. Then $\frac{d}{dt}(u\mathbf{r}) = \frac{d}{dt}[u(t)\,f(t)\,\mathbf{i} + u(t)\,g(t)\,\mathbf{j} + u(t)\,h(t)\,\mathbf{k}] = \left(\frac{du}{dt}f(t) + u(t)\frac{df}{dt}\right)\mathbf{i} + \left(\frac{du}{dt}g(t) + u(t)\frac{dg}{dt}\right)\mathbf{j} + \left(\frac{du}{dt}h(t) + u(t)\frac{dh}{dt}\right)\mathbf{k} = (f(t)\,\mathbf{i} + g(t)\,\mathbf{j} + h(t)\,\mathbf{k})\frac{du}{dt} + u(t)\left(\frac{df}{dt} + \frac{dg}{dt} + \frac{dh}{dt}\right) = \mathbf{r}\frac{du}{dt} + u\frac{d\mathbf{r}}{dt}$

12.2 MODELING PROJECTILE MOTION

1. $x = \left(v_0 \cos \alpha\right)t \Rightarrow (21 \text{ km})\left(\frac{1000 \text{ m}}{1 \text{ km}}\right) = 840 \text{ m/s}(\cos 60°)t \Rightarrow t = \frac{21\ 000 \text{ m}}{(840 \text{ km/s})(\cos 60°)} = 50 \text{ seconds}$

5. $R = \frac{v_0^2}{g} \sin 2\alpha = \frac{v_0^2}{g}(2 \sin \alpha \cos \alpha) = \frac{v_0^2}{g}(2 \cos(90° - \alpha) \sin(90° - \alpha)) = \frac{v_0^2}{g}(\sin 2(90° - \alpha))$

9. $R = \frac{v_0^2}{g} \sin 2\alpha \Rightarrow 746.4 \text{ ft}) = \frac{v_0^2}{32 \text{ ft/sec}^2} \sin 2(9°) \Rightarrow v_0^2 \approx 77\ 292.84 \text{ ft}^2/\text{sec}^2 \Rightarrow v_0 \approx 278.01 \text{ ft/sec} \approx$ 189.6 mph

13. $x = \left(v_0 \cos \alpha\right)t \Rightarrow 135 \text{ ft} = (90 \text{ ft/sec})(\cos 30°)t \Rightarrow t \approx 1.732 \text{ sec}.$ $y = \left(v_0 \sin \alpha\right)t - \frac{1}{2}gt^2 \Rightarrow$
$y \approx (90 \text{ ft/sec})(\sin 30°)(1.732 \text{ sec}) - \frac{1}{2}\left(32 \text{ ft/sec}^2\right)(1.732 \text{ sec})^2 \Rightarrow y \approx 29.94 \text{ ft}.$ The golf ball will clip the leaves at the top.

17. $x = \left(v_0 \cos \alpha\right) t \Rightarrow 315 \text{ ft} = (v_0 \cos 20°)t \Rightarrow v_0 = \frac{315}{t \cos 20°}.$ $y = \left(v_0 \sin \alpha\right)t - \frac{1}{2}gt^2 \Rightarrow$
$34 \text{ ft} = \frac{315}{t \cos 20°}(t \sin 20°) - \frac{1}{2}(32)t^2 \Rightarrow 34 = 315 \tan 20° - 16t^2 \Rightarrow t^2 \approx 5.04 \text{ sec}^2 \Rightarrow t \approx 2.25 \text{ sec}$
$t \approx 2.25 \text{ sec } \alpha v_0 = \frac{315}{(2.25)\cos 20°} \approx 148.98 \text{ ft/sec}$

21. $\frac{d\mathbf{r}}{dt} = \int (-g\,\mathbf{j})\, dt = -gt\,\mathbf{j} + \mathbf{C}_1.$ $\frac{d\mathbf{r}}{dt}(0) = (v_0 \cos \alpha)\,\mathbf{i} + (v_0 \sin \alpha)\,\mathbf{j} \Rightarrow -g(0)\,\mathbf{j} + \mathbf{C}_1 = (v_0 \cos \alpha)\,\mathbf{i} +$
$(v_0 \sin \alpha)\,\mathbf{j} \Rightarrow \mathbf{C}_1 = (v_0 \cos \alpha)\,\mathbf{i} + (v_0 \sin \alpha)\,\mathbf{j} \Rightarrow \frac{d\mathbf{r}}{dt} = (v_0 \cos \alpha)\,\mathbf{i} + (v_0 \sin \alpha - gt)\,\mathbf{j}.$
$\mathbf{r} = \int \left((v_0 \cos \alpha)\,\mathbf{i} + (v_0 \sin \alpha - gt)\,\mathbf{j}.\right) dt = (v_0 t \cos \alpha)\,\mathbf{i} + (v_0 t \sin \alpha - \frac{1}{2}gt^2)\,\mathbf{j} + \mathbf{C}_2.$
$\mathbf{r}(0) = x_0\,\mathbf{i} + y_0\,\mathbf{j} \Rightarrow (v_0(0) \cos \alpha)\,\mathbf{i} + (v_0(0) \sin \alpha - \frac{1}{2}g(0)^2)\,\mathbf{j} + \mathbf{C}_2 = x_0\,\mathbf{i} + y_0\,\mathbf{j} \Rightarrow$
$\mathbf{C}_2 = x_0\,\mathbf{i} + y_0\,\mathbf{j}.$ $\therefore$ $\mathbf{r} = (x_0 + v_0 t \cos \alpha)\,\mathbf{i} + (y_0 + v_0 t \sin \alpha - \frac{1}{2}gt^2)\,\mathbf{j} \Rightarrow x = x_0 + v_0 t \cos \alpha,$
$y = y_0 + v_0 t \sin \alpha - \frac{1}{2}gt^2$

12.3 ARC LENGTH PARAMETRIZATION

1. $\mathbf{r} = (2 \cos t)\,\mathbf{i} + (2 \sin t)\,\mathbf{j} + \sqrt{5}\,t\,\mathbf{k} \Rightarrow \mathbf{v} = (-2 \sin t)\,\mathbf{i} + (2 \cos t)\,\mathbf{j} + \sqrt{5}\,\mathbf{k} \Rightarrow$
$|\mathbf{v}| = \sqrt{(-2 \sin t)^2 + (2 \cos t)^2 + \left(\sqrt{5}\right)^2} = \sqrt{4 \sin^2 t + 4 \cos^2 t + 5} = 3.$
$\mathbf{T} = \frac{\mathbf{v}}{|\mathbf{v}|} = \left(-\frac{2}{3} \sin t\right)\mathbf{i} + \left(\frac{2}{3} \cos t\right)\mathbf{j} + \frac{\sqrt{5}}{3}\mathbf{k}.$ Length $= \int_0^{\pi} |\mathbf{v}|\, dt = \int_0^{\pi} 3\, dt = [3t]_0^{\pi} = 3\pi$

5. $\mathbf{r} = (2+t)\,\mathbf{i} - (t+1)\,\mathbf{j} + t\,\mathbf{k} \Rightarrow \mathbf{v} = \mathbf{i} - \mathbf{j} + \mathbf{k} \Rightarrow |\mathbf{v}| = \sqrt{1^2 + (-1)^2 + 1^2} = \sqrt{3}$. $\mathbf{T} = \frac{\mathbf{v}}{|\mathbf{v}|} = \frac{1}{\sqrt{3}}\mathbf{i} - \frac{1}{\sqrt{3}}\mathbf{j} + \frac{1}{\sqrt{3}}\mathbf{k}$

$$\text{Length} = \int_0^3 \sqrt{3}\,dt = \left[\sqrt{3}\,t\right]_0^3 = 3\sqrt{3}$$

9. $\mathbf{r} = (4\cos t)\,\mathbf{i} + (4\sin t)\,\mathbf{j} + 3t\,\mathbf{k} \Rightarrow \mathbf{v} = (-4\sin t)\,\mathbf{i} + (4\cos t)\,\mathbf{j} + 3\,\mathbf{k} \Rightarrow |\mathbf{v}| = \sqrt{(-4\sin t)^2 + (4\cos t)^2 + 3^2}$

$= \sqrt{25} = 5$. $s(t) = \int_0^t 5\,d\lambda = 5t$ Length $= s\left(\frac{\pi}{2}\right) = \frac{5\pi}{2}$

13. $\mathbf{r} = (\sqrt{2}\,t)\,\mathbf{i} + (\sqrt{2}\,t)\,\mathbf{j} + (1-t^2)\,\mathbf{k} \Rightarrow \mathbf{v} = \sqrt{2}\,\mathbf{i} + \sqrt{2}\,\mathbf{j} - 2t\,\mathbf{k} \Rightarrow |\mathbf{v}| = \sqrt{(\sqrt{2})^2 + (\sqrt{2})^2 + (-2t)^2} = \sqrt{4 + 4t^2}$

$= 2\sqrt{1+t^2}$ Length $= \int_0^1 2\sqrt{1+t^2}\,dt = \left[2\left(\frac{t}{2}\sqrt{1+t^2} + \frac{1}{2}\ln\left(t + \sqrt{1+t^2}\right)\right)\right]_0^1 = \sqrt{2} + \ln\left(1 + \sqrt{2}\right)$

12.4 CURVATURE, TORSION, AND THE TNB FRAME

1. $\mathbf{r} = t\,\mathbf{i} + \ln(\cos t)\,\mathbf{j} \Rightarrow \mathbf{v} = \mathbf{i} + \frac{-\sin t}{\cos t}\mathbf{j} = \mathbf{i} - \tan t\,\mathbf{j} \Rightarrow |\mathbf{v}| = \sqrt{1^2 + (-\tan t)^2} = \sqrt{\sec^2 t} = |\sec t| = \sec t$

since $-\frac{\pi}{2} < t < \frac{\pi}{2}$. $\mathbf{T} = \frac{\mathbf{v}}{|\mathbf{v}|} = \frac{1}{\sec t}\mathbf{i} - \frac{\tan t}{\sec t}\mathbf{j} = \cos t\,\mathbf{i} - \sin t\,\mathbf{j}$; $\frac{d\mathbf{T}}{dt} = -\sin t\,\mathbf{i} - \cos t\,\mathbf{j} \Rightarrow$

$\left|\frac{d\mathbf{T}}{dt}\right| = \sqrt{(-\sin t)^2 + (-\cos t)^2} = 1$. $\mathbf{N} = \frac{d\mathbf{T}/dt}{|d\mathbf{T}/dt|} = (-\sin t)\,\mathbf{i} - (\cos t)\,\mathbf{j}$. $\mathbf{a} = (-\sec^2 t)\,\mathbf{j} \Rightarrow$

$$\mathbf{v} \times \mathbf{a} = \begin{vmatrix} \mathbf{i} & \mathbf{j} & \mathbf{k} \\ 1 & -\tan t & 0 \\ 0 & -\sec^2 t & 0 \end{vmatrix} = (-\sec^2 t)\,\mathbf{k}. \quad |\mathbf{v} \times \mathbf{a}| = \sqrt{(-\sec^2 t)^2} = \sec^2 t \Rightarrow \kappa = \frac{|\mathbf{v} \times \mathbf{a}|}{|\mathbf{v}|^3} = \frac{\sec^2 t}{\sec^3 t} = \cos t$$

5. $\mathbf{r} = (3\sin t)\,\mathbf{i} + (3\cos t)\,\mathbf{j} + 4t\,\mathbf{k} \Rightarrow \mathbf{v} = (3\cos t)\,\mathbf{i} + (-3\sin t)\,\mathbf{j} + 4\,\mathbf{k} \Rightarrow |\mathbf{v}| = \sqrt{(3\cos t)^2 + (-3\sin t)^2 + 4^2}$

$= \sqrt{25} = 5$. $\mathbf{T} = \frac{\mathbf{v}}{|\mathbf{v}|} = \frac{3\cos t}{5}\mathbf{i} - \frac{3\sin t}{5}\mathbf{j} + \frac{4}{5}\mathbf{k} \Rightarrow \frac{d\mathbf{T}}{dt} = \left(-\frac{3}{5}\sin t\right)\mathbf{i} - \left(\frac{3}{5}\cos t\right)\mathbf{j}$

$\left|\frac{d\mathbf{T}}{dt}\right| = \sqrt{\left(-\frac{3}{5}\sin t\right)^2 + \left(-\frac{3}{5}\cos t\right)^2} = \frac{3}{5}$. $\mathbf{N} = \frac{d\mathbf{T}/dt}{|d\mathbf{T}/dt|} = (-\sin t)\,\mathbf{i} - (\cos t)\,\mathbf{j}$

$$\mathbf{a} = (-3\sin t)\,\mathbf{i} + (-3\cos t)\,\mathbf{j} \Rightarrow \mathbf{v} \times \mathbf{a} = \begin{vmatrix} \mathbf{i} & \mathbf{j} & \mathbf{k} \\ 3\cos t & -3\sin t & 4 \\ -3\sin t & -3\cos t & 0 \end{vmatrix} = (12\cos t)\,\mathbf{i} - (12\sin t)\,\mathbf{j} - 9\,\mathbf{k} \Rightarrow$$

5. (Continued)

$|\mathbf{v} \times \mathbf{a}| = \sqrt{(12 \cos t)^2 + (-12 \sin t)^2 + (-9)^2} = \sqrt{225} = 15.$ $\kappa = \frac{|\mathbf{v} \times \mathbf{a}|}{|\mathbf{v}|^3} = \frac{15}{5^3} = \frac{3}{25}$

$$\mathbf{B} = \mathbf{T} \times \mathbf{N} = \begin{vmatrix} \mathbf{i} & \mathbf{j} & \mathbf{k} \\ \frac{3}{5}\cos t & -\frac{3}{5}\sin t & \frac{4}{5} \\ -\sin t & -\cos t & 0 \end{vmatrix} = \left(\frac{4}{5}\cos t\right)\mathbf{i} - \left(\frac{4}{5}\sin t\right)\mathbf{j} + \left(-\frac{3}{5}\cos^2 t - \frac{3}{5}\sin^2 t\right)\mathbf{k} =$$

$$\left(\frac{4}{5}\cos t\right)\mathbf{i} - \left(\frac{4}{5}\sin t\right)\mathbf{j} - \frac{3}{5}\mathbf{k}.\ \dot{\mathbf{a}} = (-3\cos t)\,\mathbf{i} + (3\sin t)\,\mathbf{j} \Rightarrow \tau = \frac{\begin{vmatrix} 3\cos t & -3\sin t & 4 \\ -3\sin t & -3\cos t & 0 \\ -3\cos t & 3\sin t & 0 \end{vmatrix}}{|\mathbf{v} \times \mathbf{a}|^2} =$$

$$\frac{-36\sin^2 t - 36\cos^2 t}{15^2} = -\frac{4}{25}$$

9. $\mathbf{r}(t) = (t^3/3)\,\mathbf{i} + (t^2/2)\,\mathbf{j},\ t > 0 \Rightarrow \mathbf{v} = t^2\,\mathbf{i} + t\,\mathbf{j} \Rightarrow |\mathbf{v}| = \sqrt{t^4 + t^2} = t\sqrt{t^2 + 1}$ since $t > 0$. $\mathbf{T} = \frac{\mathbf{v}}{|\mathbf{v}|} = \frac{t^2\,\mathbf{i} + t\,\mathbf{j}}{t\sqrt{t^2+1}} =$

$$\frac{t}{\sqrt{t^2+1}}\mathbf{i} + \frac{\mathbf{j}}{\sqrt{t^2+1}} \Rightarrow \frac{d\mathbf{T}}{dt} = \frac{\mathbf{i}}{\left(t^2+1\right)^{3/2}} - \frac{t\,\mathbf{j}}{\left(t^2+1\right)^{3/2}} \Rightarrow \left|\frac{d\mathbf{T}}{dt}\right| = \sqrt{\left(\frac{1}{\left(t^2+1\right)^{3/2}}\right)^2 + \left(\frac{-t}{\left(t^2+1\right)^{3/2}}\right)^2} =$$

$$\sqrt{\frac{1+t^2}{\left(t^2+1\right)^3}} = \frac{1}{t^2+1}.\ \mathbf{N} = \frac{d\mathbf{T}/dt}{|d\mathbf{T}/dt|} = \frac{\mathbf{i}}{\sqrt{t^2+1}} - \frac{t\,\mathbf{j}}{\sqrt{t^2+1}}$$

$$\mathbf{a} = 2t\,\mathbf{i} + \mathbf{j} \Rightarrow \mathbf{v} \times \mathbf{a} = \begin{vmatrix} \mathbf{i} & \mathbf{j} & \mathbf{k} \\ t^2 & t & 0 \\ 2t & 1 & 0 \end{vmatrix} = -t^2\,\mathbf{k} \Rightarrow |\mathbf{v} \times \mathbf{a}| = \sqrt{(-t^2)^2} = t^2.\ \therefore\ \kappa = \frac{|\mathbf{v} \times \mathbf{a}|}{|\mathbf{v}|^3} = \frac{t^2}{\left(t\sqrt{t^2+1}\right)^3} =$$

$$\frac{1}{t\left(t^2+1\right)^{3/2}}.\ \mathbf{B} = \mathbf{T} \times \mathbf{N} = \begin{vmatrix} \mathbf{i} & \mathbf{j} & \mathbf{k} \\ \frac{t}{\sqrt{t^2+1}} & \frac{1}{\sqrt{t^2+1}} & 0 \\ \frac{1}{\sqrt{t^2+1}} & \frac{-t}{\sqrt{t^2+1}} & 0 \end{vmatrix} = -\mathbf{k}.\quad \dot{\mathbf{a}} = 2\,\mathbf{i} \Rightarrow \tau = \frac{\begin{vmatrix} t^2 & t & 0 \\ 2t & 1 & 0 \\ 2 & 0 & 0 \end{vmatrix}}{|\mathbf{v} \times \mathbf{a}|^2} = 0$$

13. $\mathbf{r} = (2t+3)\,\mathbf{i} + (t^2 - 1)\,\mathbf{j} \Rightarrow \mathbf{v} = 2\,\mathbf{i} + 2t\,\mathbf{j} \Rightarrow |\mathbf{v}| = \sqrt{2^2 + (2t)^2} = 2\sqrt{1+t^2}$

$$a_T = 2\left(\frac{1}{2}\right)(1+t^2)^{-1/2}(2t) = \frac{2t}{\sqrt{1+t^2}};\ \mathbf{a} = 2\,\mathbf{j} \Rightarrow |\mathbf{a}| = 2 \Rightarrow a_N = \sqrt{|\mathbf{a}|^2 - {a_T}^2} = \sqrt{2^2 - \left(\frac{2t}{\sqrt{1+t^2}}\right)^2}$$

$$= \frac{2}{\sqrt{1+t^2}}.\ \therefore\ \mathbf{a} = \frac{2t}{\sqrt{1+t^2}}\mathbf{T} + \frac{2}{\sqrt{1+t^2}}\mathbf{N}.$$

17. $\mathbf{r} = (t+1)\,\mathbf{i} + 2t\,\mathbf{j} + t^2\,\mathbf{k} \Rightarrow \mathbf{v} = \mathbf{i} + 2\,\mathbf{j} + 2t\,\mathbf{k} \Rightarrow |\mathbf{v}| = \sqrt{1^2 + 2^2 + (2t)^2} =$

$$\sqrt{5+4t^2}.\ a_T = \frac{1}{2}\left(5+4t^2\right)^{-1/2}(8t) = 4t\left(5+4t^2\right)^{-1/2} \Rightarrow a_T(1) = \frac{4}{\sqrt{9}} = \frac{4}{3}.\ \mathbf{a} = 2\,\mathbf{k} \Rightarrow \mathbf{a}(1) = 2\,\mathbf{k} \Rightarrow$$

$$|\mathbf{a}(1)| = 2.\ a_N = \sqrt{|\mathbf{a}|^2 - {a_T}^2} = \sqrt{2^2 - \left(\frac{4}{3}\right)^2} = \frac{2\sqrt{5}}{3}.\ \therefore\ \mathbf{a}(1) = \frac{4}{3}\mathbf{T} + \frac{2\sqrt{5}}{3}\mathbf{N}$$

21. $\mathbf{r} = (\cos t)\,\mathbf{i} + (\sin t)\,\mathbf{j} - \mathbf{k} \Rightarrow \mathbf{v} = (-\sin t)\,\mathbf{i} + (\cos t)\,\mathbf{j} \Rightarrow |\mathbf{v}| = \sqrt{(-\sin t)^2 + (\cos t)^2} = 1$. $\mathbf{T} = \frac{\mathbf{v}}{|\mathbf{v}|} =$ $(-\sin t)\,\mathbf{i} + (\cos t)\,\mathbf{j} \Rightarrow \mathbf{T}\left(\frac{\pi}{4}\right) = -\frac{\sqrt{2}}{2}\mathbf{i} + \frac{\sqrt{2}}{2}\mathbf{j}$. $\frac{d\mathbf{T}}{dt} = (-\cos t)\,\mathbf{i} - (\sin t)\,\mathbf{j} \Rightarrow \left|\frac{d\mathbf{T}}{dt}\right| = \sqrt{(-\cos t)^2 + (-\sin t)^2}$ $= 1$. $\mathbf{N} = \frac{d\mathbf{T}/dt}{|d\mathbf{T}/dt|} = (-\cos t)\,\mathbf{i} - (\sin t)\,\mathbf{j} \Rightarrow \mathbf{N}\left(\frac{\pi}{4}\right) = -\frac{\sqrt{2}}{2}\mathbf{i} - \frac{\sqrt{2}}{2}\mathbf{j}$. $\mathbf{r}\left(\frac{\pi}{4}\right) = \frac{\sqrt{2}}{2}\mathbf{i} + \frac{\sqrt{2}}{2}\mathbf{j} - \mathbf{k}$

$$\mathbf{B} = \mathbf{T} \times \mathbf{N} = \begin{vmatrix} \mathbf{i} & \mathbf{j} & \mathbf{k} \\ -\sin t & \cos t & 0 \\ -\cos t & -\sin t & 0 \end{vmatrix} = \mathbf{k} \Rightarrow \mathbf{B}\left(\frac{\pi}{4}\right) = \mathbf{k}.\ P = \left(\frac{\sqrt{2}}{2}, \frac{\sqrt{2}}{2}, -1\right)\left(\text{see } \mathbf{r}\left(\frac{\pi}{4}\right)\right)$$, the osculating plane is $z = -1$ since $\mathbf{B}$ is the normal vector and $(0)x + (0)y + (1)z = (0)\left(\frac{\sqrt{2}}{2}\right) + (0)\left(\frac{\sqrt{2}}{2}\right) + (1)(-1)$.

The normal plane is $-x + y = 0$ since $\mathbf{T}$ is the normal vector and $-\frac{\sqrt{2}}{2}x + \frac{\sqrt{2}}{2}y + (0)z = \left(-\frac{\sqrt{2}}{2}\right)\left(\frac{\sqrt{2}}{2}\right) + \left(\frac{\sqrt{2}}{2}\right)\left(\frac{\sqrt{2}}{2}\right) + (-1)(0) \Rightarrow -\frac{\sqrt{2}}{2}x + \frac{\sqrt{2}}{2}y = 0$. The rectifying plane is $x + y = \sqrt{2}$ since $\mathbf{N}$ is the normal vector and $-\frac{\sqrt{2}}{2}x - \frac{\sqrt{2}}{2}y + (0)z = \left(-\frac{\sqrt{2}}{2}\right)\left(\frac{\sqrt{2}}{2}\right) - \left(\frac{\sqrt{2}}{2}\right)\left(\frac{\sqrt{2}}{2}\right) + (-1)(0) \Rightarrow -\frac{\sqrt{2}}{2}x - \frac{\sqrt{2}}{2}y = -1$.

25. If acceleration is perpendicular to the velocity, then $a_T = 0 \Rightarrow |\mathbf{v}|$ is constant.

29. $\kappa = \frac{a}{a^2 + b^2} \Rightarrow \frac{d\kappa}{da} = \frac{-a^2 + b^2}{(a^2 + b^2)^2}$. If $\frac{d\kappa}{da} = 0$, then $-a^2 + b^2 = 0 \Rightarrow a = \pm b$. When $a = b$ ($b > 0$), $\frac{d\kappa}{da} > 0$ if $a < b$ and $\frac{d\kappa}{da} < 0$ if $a > b$. $\therefore$ κ is at a maximum when $a = b \Rightarrow \kappa(b) = \frac{b}{b^2 + b^2} = \frac{1}{2b}$, the maximum value of κ.

33. $\mathbf{r} = t\,\mathbf{i} + (\sin t)\,\mathbf{j} \Rightarrow \mathbf{v} = \mathbf{i} + (\cos t)\,\mathbf{j} \Rightarrow |\mathbf{v}| = \sqrt{1^2 + (\cos t)^2} = \sqrt{1 + \cos^2 t} \Rightarrow \left|\mathbf{v}\left(\frac{\pi}{2}\right)\right| = \sqrt{1 + \cos^2\left(\frac{\pi}{2}\right)}$

$$= 1.\ \mathbf{a} = (-\sin t)\,\mathbf{j} \Rightarrow \mathbf{v} \times \mathbf{a} = \begin{vmatrix} \mathbf{i} & \mathbf{j} & \mathbf{k} \\ 1 & \cos t & 0 \\ 0 & -\sin t & 0 \end{vmatrix} = (-\sin t)\,\mathbf{k} \Rightarrow |\mathbf{v} \times \mathbf{a}| = \sqrt{(-\sin t)^2} = |\sin t| \Rightarrow$$

$|\mathbf{v} \times \mathbf{a}|\left(\frac{\pi}{2}\right) = \left|\sin\left(\frac{\pi}{2}\right)\right| = 1 \Rightarrow \kappa = \frac{|\mathbf{v} \times \mathbf{a}|}{|\mathbf{v}|^3} = \frac{1}{1^3} = 1$. $\therefore$ $\rho = \frac{1}{1} = 1 \Rightarrow$ center is $\left(\frac{\pi}{2}, 0\right)$, $r = 1 \Rightarrow$ $\left(x - \frac{\pi}{2}\right)^2 + y^2 = 1$

37. $y = x^2 \Rightarrow f'(x) = 2x$, $f''(x) = 2 \Rightarrow \kappa = \frac{|2|}{\left(1 + (2x)^2\right)^{3/2}} = \frac{2}{\left(1 + 4x^2\right)^{3/2}}$ (Compare with κ in Exercise 32 b, with $x = t$.)

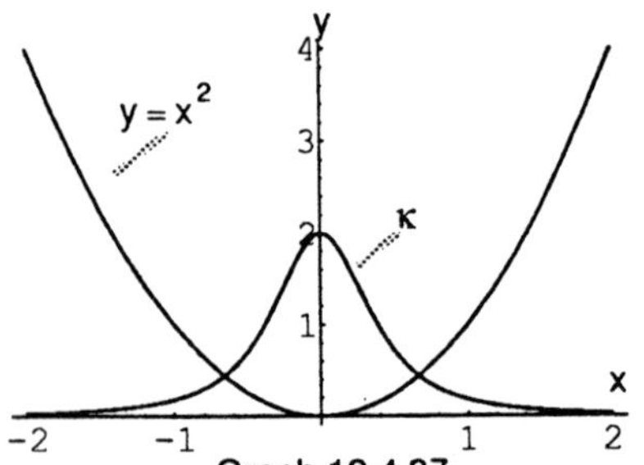

Graph 12.4.37

12.5 PLANETARY MOTION AND SATELLITES

1. $\frac{T^2}{a^3} = \frac{4\pi^2}{GM} \Rightarrow T^2 = \frac{4\pi^2}{GM} a^3 \Rightarrow T^2 = \frac{4\pi^2}{(6.6720 \times 10^{-11} Nm^2kg^{-2})(5.975 \times 10^{24} kg)} (6\,808\,000 \text{ m})^3 \approx$ $3.125 \times 10^7 \sec^2 \Rightarrow T \approx \sqrt{3125 \times 10^4 \sec^2} \approx 55.90 \times 10^2 \sec \approx 93.17$ minutes.

5. $a = 22030$ km $= 2.203 \times 10^7$ m. $T^2 = \frac{4\pi^2}{GM} a^3 \Rightarrow T^2 =$ $\frac{4\pi^2}{(6.670 \times 10^{-11} Nm^2kg^{-2})(6.418 \times 10^{23} kg)} (2.203 \times 10^7 s)^3 \approx 9.857 \times 10^9 \sec^2 \Rightarrow T \approx$ $\sqrt{9.857 \times 10^8 \sec^2} \approx 9.928 \times 10^4 \sec \approx 1655$ minutes.

9. $r = \frac{GM}{v^2} \Rightarrow v^2 = \frac{GM}{r} \Rightarrow |v| = \sqrt{\frac{GM}{r}} = \sqrt{\frac{(6.6720 \times 10^{-11} Nm^2kg^{-2})(5.975 \times 10^{24} kg)}{r}}$ $\approx 1.9966 \times 10^7 r^{-1/2}$ m/s

12.P PRACTICE EXERCISES

1. $\mathbf{r} = (4 \cos t)\,\mathbf{i} + (\sqrt{2} \sin t)\,\mathbf{j} \Rightarrow x = 4 \cos t \Rightarrow x^2 = 16 \cos^2 t$

$y = \sqrt{2} \sin t \Rightarrow y^2 = 2 \sin^2 t \Rightarrow 8y^2 = 16 \sin^2 t \Rightarrow$
$x^2 + 8y^2 = 16 \Rightarrow \frac{x^2}{16} + \frac{y^2}{2} = 1$. $t = 0 \Rightarrow x = 4, y = 0$;
$t = \frac{\pi}{4} \Rightarrow x = 2\sqrt{2}, y = 1$. $\mathbf{v} = (-4 \sin t)\,\mathbf{i} + (\sqrt{2} \cos t)\,\mathbf{j}$
$\Rightarrow \mathbf{v}(0) = \sqrt{2}\,\mathbf{j}, \mathbf{v}\left(\frac{\pi}{4}\right) = -2\sqrt{2}\,\mathbf{i} + \mathbf{j}$.
$\mathbf{a} = (-4 \cos t)\,\mathbf{i} + (-\sqrt{2} \sin t)\,\mathbf{j} \Rightarrow \mathbf{a}(0) = -4\,\mathbf{i}, \mathbf{a}\left(\frac{\pi}{4}\right) =$
$-2\sqrt{2}\,\mathbf{i} - \mathbf{j}$.

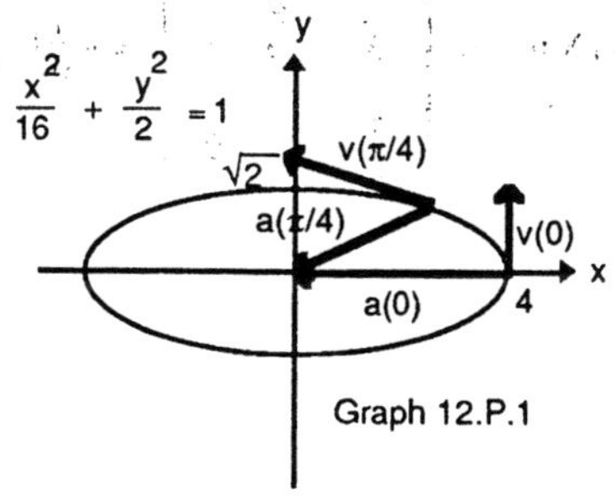

Graph 12.P.1

5. $\mathbf{r} = \int((-\sin t)\,\mathbf{i} + (\cos t)\,\mathbf{j} + \mathbf{k})\,dt = (\cos t)\,\mathbf{i} + (\sin t)\,\mathbf{j} + t\,\mathbf{k} + \mathbf{C}$. $\mathbf{r}(0) = \mathbf{j} \Rightarrow (\cos 0)\,\mathbf{i} + (\sin 0)\,\mathbf{j} + (0)\,\mathbf{k}$ $+ \mathbf{C} = \mathbf{j} \Rightarrow \mathbf{C} = \mathbf{j} - \mathbf{i} \Rightarrow \mathbf{r} = ((\cos t) - 1)\,\mathbf{i} + ((\sin t) + 1)\,\mathbf{j} + t\,\mathbf{k}$

9. $\mathbf{r} = (2 \cos t)\,\mathbf{i} + (2 \sin t)\,\mathbf{j} + t^2\,\mathbf{k} \Rightarrow \mathbf{v} = (-2 \sin t)\,\mathbf{i} + (2 \cos t)\,\mathbf{j} + 2t\,\mathbf{k} \Rightarrow$

$|\mathbf{v}| = \sqrt{(-2 \sin t)^2 + (2 \cos t)^2 + (2t)^2} = 2\sqrt{1 + t^2}$. Length $= \int_0^{\pi/4} 2\sqrt{1 + t^2}\,dt =$

$\left[t\sqrt{1 + t^2} + \ln\left|t + \sqrt{1 + t^2}\right|\right]_0^{\pi/4} = \frac{\pi}{4}\sqrt{1 + \frac{\pi^2}{16}} + \ln\left(\frac{\pi}{4} + \sqrt{1 + \frac{\pi^2}{16}}\right)$

13. $\mathbf{r} = t\,\mathbf{i} + \frac{1}{2}e^{2t}\,\mathbf{j},\ t = \ln 2 \Rightarrow \mathbf{v} = \mathbf{i} + e^{2t}\,\mathbf{j} \Rightarrow |\mathbf{v}| = \sqrt{1 + e^{4t}} \Rightarrow \mathbf{T} = \frac{\mathbf{v}}{|\mathbf{v}|} = \frac{\mathbf{i} + e^{2t}\,\mathbf{j}}{\sqrt{1+e^{4t}}} = \frac{1}{\sqrt{1+e^{4t}}}\mathbf{i} + \frac{e^{2t}}{\sqrt{1+e^{4t}}}\mathbf{j}$

$\Rightarrow \frac{d\mathbf{T}}{dt} = \frac{-2e^{4t}}{(1+e^{4t})^{3/2}}\mathbf{i} + \frac{2e^{2t}}{(1+e^{4t})^{3/2}}\mathbf{j} \Rightarrow \left|\frac{d\mathbf{T}}{dt}\right| = \sqrt{\frac{4e^{8t} + 4e^{4t}}{(1+e^{4t})^3}} = \frac{2e^{2t}}{1+e^{4t}} \Rightarrow \mathbf{N} = \frac{d\mathbf{T}/dt}{|d\mathbf{T}/dt|} = \frac{-e^{2t}}{(1+e^{4t})^{1/2}}\mathbf{i} +$
$\frac{1}{(1+e^{4t})^{1/2}}\mathbf{j}$. $\therefore\ \mathbf{T}(\ln 2) = \frac{1}{\sqrt{17}}\mathbf{i} + \frac{4}{\sqrt{17}}\mathbf{j}$, $\mathbf{N}(\ln 2) = -\frac{4}{\sqrt{17}}\mathbf{i} + \frac{1}{\sqrt{17}}\mathbf{j}$. $\mathbf{B}(\ln 2) = \mathbf{T}(\ln 2) \times \mathbf{N}(\ln 2) =$

$$\begin{vmatrix} \mathbf{i} & \mathbf{j} & \mathbf{k} \\ \frac{1}{\sqrt{17}} & \frac{4}{\sqrt{17}} & 0 \\ -\frac{4}{\sqrt{17}} & \frac{1}{\sqrt{17}} & 0 \end{vmatrix} = \mathbf{k}.\ \mathbf{a} = 2e^{2t}\,\mathbf{j} \Rightarrow \mathbf{a}(\ln 2) = 8\,\mathbf{j}.\ \mathbf{v}(\ln 2) = \mathbf{i} + 4\,\mathbf{j} \Rightarrow \mathbf{v}(\ln 2) \times \mathbf{a}(\ln 2) = \begin{vmatrix} \mathbf{i} & \mathbf{j} & \mathbf{k} \\ 1 & 4 & 0 \\ 0 & 8 & 0 \end{vmatrix}$$

$= 8\,\mathbf{k} \Rightarrow |\mathbf{v}(\ln 2) \times \mathbf{a}(\ln 2)| = 8.\ |\mathbf{v}(\ln 2)| = \sqrt{17}. \Rightarrow \kappa = \frac{8}{(\sqrt{17})^3} = \frac{8}{17\sqrt{17}}.\ \dot{\mathbf{a}} = 4e^{2t}\,\mathbf{j} \Rightarrow \dot{\mathbf{a}}(\ln 2) = 16\,\mathbf{j} \Rightarrow$

$$\tau = \frac{\begin{vmatrix} 1 & 4 & 0 \\ 0 & 8 & 0 \\ 0 & 16 & 0 \end{vmatrix}}{|\mathbf{v} \times \mathbf{a}|^2} = 0.$$

$t = \ln 2 \Rightarrow (\ln 2, 2, 0)$ is on the curve.

17. $\mathbf{r} = (\sin t)\,\mathbf{i} + (\sqrt{2}\cos t)\,\mathbf{j} + (\sin t)\,\mathbf{k} \Rightarrow \mathbf{v} = (\cos t)\,\mathbf{i} - (\sqrt{2}\sin t)\,\mathbf{j} + (\cos t)\,\mathbf{k} \Rightarrow$
$|\mathbf{v}| = \sqrt{(\cos t)^2 + (-\sqrt{2}\sin t)^2 + (\cos t)^2} = \sqrt{2} \Rightarrow \mathbf{T} = \frac{\mathbf{v}}{|\mathbf{v}|} = \frac{(\cos t)\,\mathbf{i} - (\sqrt{2}\sin t)\,\mathbf{j} + (\cos t)\,\mathbf{k}}{\sqrt{2}} =$
$\left(\frac{1}{\sqrt{2}}\cos t\right)\mathbf{i} - (\sin t)\,\mathbf{j} + \left(\frac{1}{\sqrt{2}}\cos t\right)\mathbf{k}.\ \frac{d\mathbf{T}}{dt} = \left(-\frac{1}{\sqrt{2}}\sin t\right)\mathbf{i} - (\cos t)\,\mathbf{j} - \left(\frac{1}{\sqrt{2}}\sin t\right)\mathbf{k} \Rightarrow$
$\left|\frac{d\mathbf{T}}{dt}\right| = \sqrt{\left(-\frac{1}{\sqrt{2}}\sin t\right)^2 + (-\cos t)^2 + \left(-\frac{1}{\sqrt{2}}\sin t\right)^2} = 1.\ \mathbf{N} = \frac{d\mathbf{T}/dt}{|d\mathbf{T}/dt|} = \left(-\frac{1}{\sqrt{2}}\sin t\right)\mathbf{i} - (\cos t)\,\mathbf{j} -$

$$\left(\frac{1}{\sqrt{2}}\sin t\right)\mathbf{k}.\ \mathbf{B} = \mathbf{T} \times \mathbf{N} = \begin{vmatrix} \mathbf{i} & \mathbf{j} & \mathbf{k} \\ \frac{1}{\sqrt{2}}\cos t & -\sin t & \frac{1}{\sqrt{2}}\cos t \\ -\frac{1}{\sqrt{2}}\sin t & -\cos t & -\frac{1}{\sqrt{2}}\sin t \end{vmatrix} = \frac{1}{\sqrt{2}}\mathbf{i} - \frac{1}{\sqrt{2}}\mathbf{k}$$

$$\mathbf{a} = (-\sin t)\,\mathbf{i} - (\sqrt{2}\cos t)\,\mathbf{j} - (\sin t)\,\mathbf{k} \Rightarrow \mathbf{v} \times \mathbf{a} = \begin{vmatrix} \mathbf{i} & \mathbf{j} & \mathbf{k} \\ \cos t & -\sqrt{2}\sin t & \cos t \\ -\sin t & -\sqrt{2}\cos t & -\sin t \end{vmatrix} = \sqrt{2}\,\mathbf{i} - \sqrt{2}\,\mathbf{k}$$

$|\mathbf{v} \times \mathbf{a}| = \sqrt{(\sqrt{2})^2 + (-\sqrt{2})^2} = \sqrt{4} = 2 \Rightarrow \kappa = \frac{|\mathbf{v} \times \mathbf{a}|}{|\mathbf{v}|^3} = \frac{2}{(\sqrt{2})^3} = \frac{1}{\sqrt{2}}$.

$$\dot{\mathbf{a}} = (-\cos t)\,\mathbf{i} + (\sqrt{2}\sin t)\,\mathbf{j} - (\cos t)\,\mathbf{k} \Rightarrow \tau = \frac{\begin{vmatrix} \cos t & -\sqrt{2}\sin t & \cos t \\ -\sin t & -\sqrt{2}\cos t & -\sin t \\ -\cos t & \sqrt{2}\sin t & -\cos t \end{vmatrix}}{|\mathbf{v} \times \mathbf{a}|^2} = = \frac{0}{|\mathbf{v} \times \mathbf{a}|^2} = 0$$

21. $\mathbf{r} = 2\mathbf{i} + \left(4\sin\frac{t}{2}\right)\mathbf{j} + \left(3 - \frac{t}{\pi}\right)\mathbf{k} \Rightarrow \mathbf{r}\cdot(\mathbf{i}-\mathbf{j}) = 2(1) + \left(4\sin\frac{t}{2}\right)(-1)$. $\mathbf{r}\cdot(\mathbf{i}-\mathbf{j}) = 0 \Rightarrow 2 - 4\sin\frac{t}{2} = 0 \Rightarrow$ $\sin\frac{t}{2} = \frac{1}{2} \Rightarrow \frac{t}{2} = \frac{\pi}{6} \Rightarrow t = \frac{\pi}{3}$ (for the first time).

25. $R = \frac{v_0^2}{g}\sin 2\alpha \Rightarrow 109.5 \text{ ft} = \frac{v_0^2}{32 \text{ ft/sec}^2}(\sin 2(45°)) \Rightarrow v_0^2 = 3504 \text{ ft}^2/\text{sec}^2 \Rightarrow v_0 = \sqrt{3504 \text{ ft}^2/\text{sec}^2}$ ≈ 59.19 ft/sec

29. $\mathbf{v} = 3\mathbf{i} + 4\mathbf{j}, \mathbf{a} = 5\mathbf{i} + 15\mathbf{j} \Rightarrow \mathbf{v} \times \mathbf{a} = \begin{vmatrix} \mathbf{i} & \mathbf{j} & \mathbf{k} \\ 3 & 4 & 0 \\ 5 & 15 & 0 \end{vmatrix} = 25\mathbf{k} \Rightarrow |\mathbf{v} \times \mathbf{a}| = \sqrt{25^2} = 25.$ $|\mathbf{v}| = \sqrt{3^2 + 4^2}$ $= 5.$ $\therefore \kappa = \frac{|\mathbf{v} \times \mathbf{a}|}{|\mathbf{v}|^3} = \frac{25}{5^3} = \frac{1}{5}.$

33. a) Given $f(x) = x - 1 - \frac{1}{2}\sin x = 0$, $f(0) = -1$ and $f(2) = 2 - 1 - \frac{1}{2}\sin 2 \geq \frac{1}{2}$ since $|\sin 2| \leq 1$. Since f is continuous on [0,2], the intermediate value theorem implies there is a root between 0 and 2.
b) Root ≈ 1.49870113

CHAPTER 13

PARTIAL DERIVATIVES

13.1 FUNCTIONS OF SEVERAL INDEPENDENT VARIABLES

1. Domain: All points in the xy–plane; Range: All Real Numbers
   Level curves are straight lines parallel to the line $y = x$.

5. Domain: All points in the xy–plane; Range: All Real Numbers
   Level curves are hyperbolas with the x- and y-axes as asymptotes when $f(x,y) \neq 0$, and the x- and y-axes when $f(x,y) = 0$.

9. a)

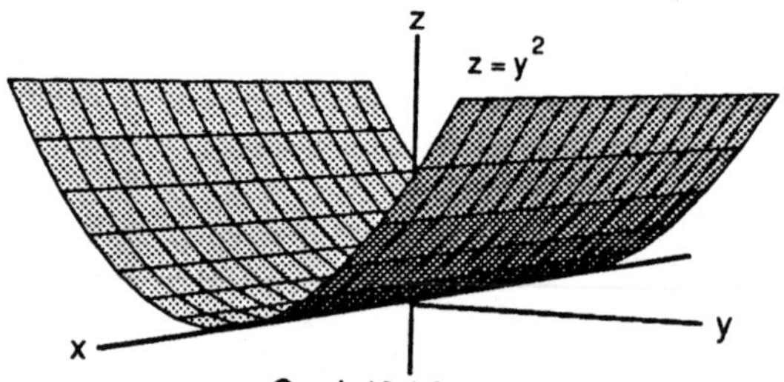

Graph 13.1.9 a

b)

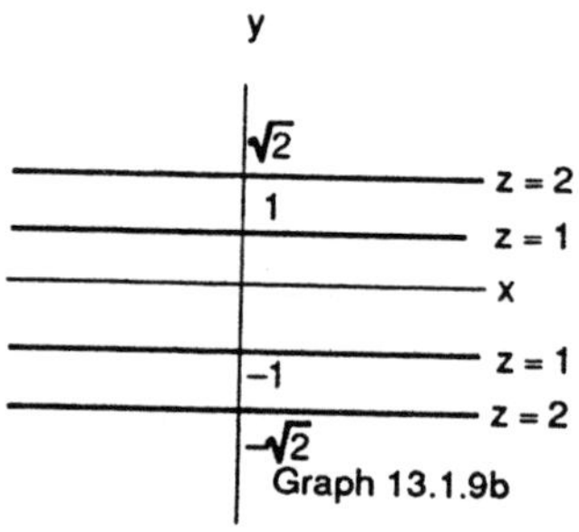

Graph 13.1.9b

13. a)

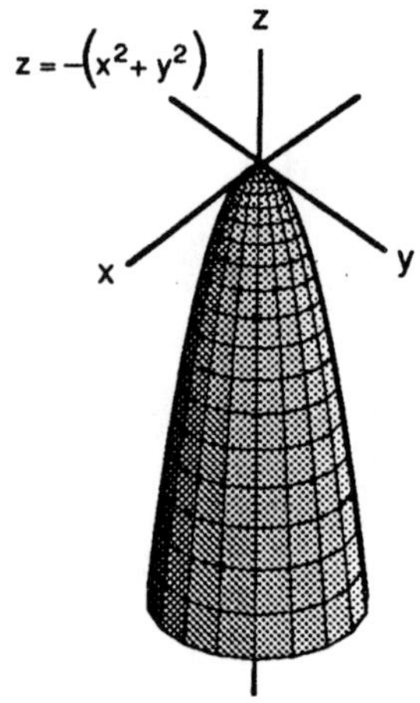

Graph 13.1.13 a

b)

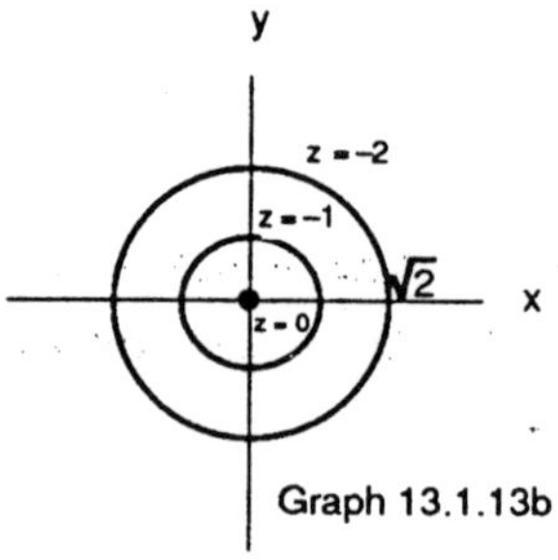

Graph 13.1.13b

17. f

21. d

25.

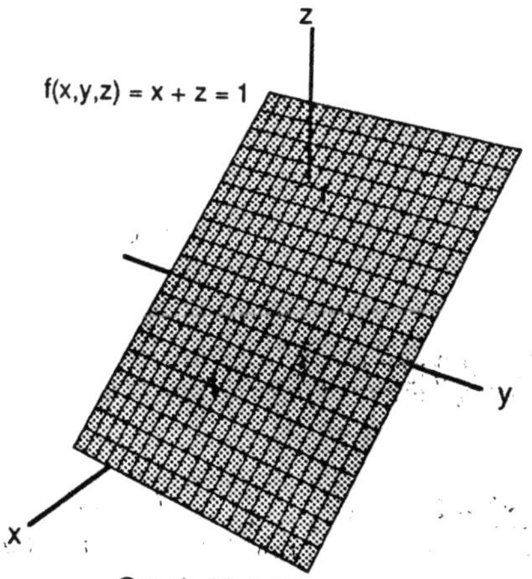

Graph 13.1.25

29.

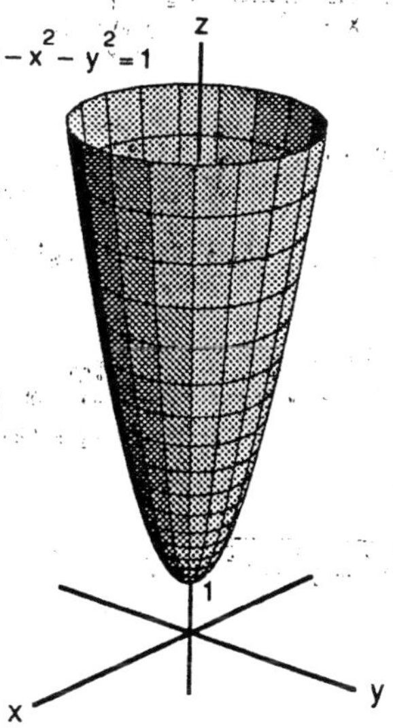

Graph 13.1.29

33. $f(x,y,z) = \sqrt{x-y} - \ln z$ at $(3,-1,1) \Rightarrow w = \sqrt{x-y} - \ln z$. At $(3,-1,1)$, $w = \sqrt{3-(-1)} - \ln 1 = 2$.
$\therefore 2 = \sqrt{x-y} - \ln z \Rightarrow \ln z = \sqrt{x-y} - 2 \Rightarrow z = e^{\sqrt{x-y} - 2}$

37. $w = 4\left(\frac{Th}{d}\right)^{1/2} = 4\left(\frac{(290\ k)(16.8\ km)}{5\ k/km}\right)^{1/2} = 124.86$ km. $\therefore$ must be 62.43 km south of Nantucket.

13.2 LIMITS AND CONTINUITY

1. $\lim_{(x,y)\to(0,0)} \frac{3x^2 - y^2 + 5}{x^2 + y^2 + 2} = \frac{5}{2}$

5. $\lim_{(x,y)\to(3,4)} \sqrt{x^2 + y^2 - 1} = \sqrt{24} = 2\sqrt{6}$

9. $\lim_{(x,y)\to(1,1)} \cos\left(\sqrt[3]{|xy| - 1}\right) = 1$

13. $\lim_{\substack{(x,y)\to(1,1)\\ x \neq y}} \frac{x^2 - 2xy + y^2}{x - y} = \lim_{(x,y)\to(1,1)} \frac{(x-y)^2}{x - y} = \lim_{(x,y)\to(1,1)} (x - y) = 0$

17. $\lim_{(x,y)\to(0,0)} \frac{x - y + 2\sqrt{x} - 2\sqrt{y}}{\sqrt{x} - \sqrt{y}} = \lim_{(x,y)\to(0,0)} \frac{\left(\sqrt{x} - \sqrt{y}\right)\left(\sqrt{x} + \sqrt{y} + 2\right)}{\sqrt{x} - \sqrt{y}} = \lim_{(x,y)\to(0,0)} \left(\sqrt{x} + \sqrt{y} + 2\right) = 2.$

Note: (x,y) must approach (0,0) through the first quadrant onlyx, $x \neq y$.

21. $\lim_{P\to(2,3,-6)} \sqrt{x^2 + y^2 + z^2} = 7$

25. $\lim_{P\to(-1/4,\pi/2,2)} \tan^{-1}(xyz) = \tan^{-1}\left(-\frac{\pi}{4}\right)$

29. a) Continuous at all (x,y) except where x = 0 or y = 0

b) Continuous at all (x,y)

33. a) Continuous at all (x,y,z) so that $(x,y,z) \neq (x,y,0)$

b) Continuous at all (x,y,z) except those on the cylinder $x^2 + z^2 = 1$

37. $\lim\limits_{\substack{(x,y)\to(0,0) \\ \text{along } y = kx^2}} \dfrac{x^4 - y^2}{x^4 + y^2} = \lim\limits_{x\to 0} \dfrac{x^4 - (kx^2)^2}{x^4 + (kx^2)^2} = \lim\limits_{x\to 0} \dfrac{x^4 - k^2x^4}{x^4 + k^2x^4} = \dfrac{1 - k^2}{1 + k^2} \Rightarrow$ different limits for

different values of k. ∴ consider paths along $y = kx^2$, k a constant.

41. $\lim\limits_{\substack{(x,y)\to(0,0) \\ \text{along } y = kx^2 \\ k \neq 0}} \dfrac{x^2 + y}{y} = \lim\limits_{x\to 0} \dfrac{x^2 + kx^2}{kx^2} = \dfrac{1 + k}{k} \Rightarrow$ different limits for different values of k. ∴ consider

paths along $y = kx^2$, k a constant, $k \neq 0$.

45. $\lim\limits_{(x,y)\to(0,0)} 1 - \dfrac{x^2y^2}{3} = 1, \quad \lim\limits_{(x,y)\to(0,0)} 1 = 1 \Rightarrow \lim\limits_{(x,y)\to(0,0)} \dfrac{\tan^{-1}xy}{xy} = 1$, by the Sandwich Theorem.

49. a) $f(x,y)\big|_{y = mx} = \dfrac{2m}{1 + m^2} = \dfrac{2\tan\theta}{1 + \tan^2\theta} = \sin 2\theta$. The value of f(x,y) is $\sin 2\theta$ where $\tan\theta = m$ along $y = mx$.

b) Since $f(x,y)\big|_{y = mx} = \sin 2\theta$ and since $-1 \leq \sin 2\theta \leq 1$ for every θ, $\lim\limits_{(x,y)\to(0,0)} f(x,y)$ varies from −1 to 1 along $y = mx$.

13.3 PARTIAL DERIVATIVES

1. $\dfrac{\partial f}{\partial x} = 2, \dfrac{\partial f}{\partial y} = 0$

5. $\dfrac{\partial f}{\partial x} = y - 1, \dfrac{\partial f}{\partial y} = x$

9. $\dfrac{\partial f}{\partial x} = 2x(5y - 1)^3$, $\dfrac{\partial f}{\partial y} = 15x^2(5y - 1)^2$

13. $\dfrac{\partial f}{\partial x} = \dfrac{-y^2 - 1}{(xy - 1)^2}, \dfrac{\partial f}{\partial y} = \dfrac{-x^2 - 1}{(xy - 1)^2}$

17. $\dfrac{\partial f}{\partial x} = e^x \sin(y + 1), \dfrac{\partial f}{\partial y} = e^x \cos(y + 1)$

21. $f_x(x,y,z) = -x(x^2 + y^2 + z^2)^{-3/2}$, $f_y(x,y,z) = -y(x^2 + y^2 + z^2)^{-3/2}$, $f_z(x,y,z) = -z(x^2 + y^2 + z^2)^{-3/2}$

25. $\dfrac{\partial f}{\partial t} = -2\pi \sin(2\pi t - \alpha), \dfrac{\partial f}{\partial \alpha} = \sin(2\pi t - \alpha)$

29. $\dfrac{\partial W}{\partial P} = V, \dfrac{\partial W}{\partial V} = P + \dfrac{\delta v^2}{2g}, \dfrac{\partial W}{\partial \delta} = \dfrac{Vv^2}{2g}, \dfrac{\partial W}{\partial v} = \dfrac{V\delta v}{g}, \dfrac{\partial W}{\partial g} = -\dfrac{V\delta v^2}{2g^2}$

33. $\dfrac{\partial g}{\partial x} = 2xy + y\cos x, \dfrac{\partial g}{\partial y} = x^2 - \sin y + \sin x, \dfrac{\partial^2 g}{\partial x^2} = 2y - y\sin x, \dfrac{\partial^2 g}{\partial y^2} = -\cos y, \dfrac{\partial^2 g}{\partial y\partial x} = \dfrac{\partial^2 g}{\partial x\partial y} = 2x + \cos x$

37. $\dfrac{\partial w}{\partial x} = \dfrac{2}{2x + 3y}, \dfrac{\partial w}{\partial y} = \dfrac{3}{2x + 3y}, \dfrac{\partial^2 w}{\partial y\partial x} = \dfrac{-6}{(2x + 3y)^2}$ and $\dfrac{\partial^2 w}{\partial x\partial y} = \dfrac{-6}{(2x + 3y)^2}$

41. a) x first b) y first c) x first d) x first e) y first f) y first

45. $\frac{\partial f}{\partial x} = 2x, \frac{\partial f}{\partial y} = 2y, \frac{\partial f}{\partial z} = -4z \Rightarrow \frac{\partial^2 f}{\partial x^2} = 2, \frac{\partial^2 f}{\partial y^2} = 2, \frac{\partial^2 f}{\partial z^2} = -4 \Rightarrow \frac{\partial^2 f}{\partial x^2} + \frac{\partial^2 f}{\partial y^2} + \frac{\partial^2 f}{\partial z^2} = 2 + 2 + (-4) = 0$

49. $\frac{\partial f}{\partial x} = -\frac{1}{2}(x^2 + y^2 + z^2)^{-3/2}(2x) = -x(x^2 + y^2 + x^2)^{-3/2}, \frac{\partial f}{\partial y} = -\frac{1}{2}(x^2 + y^2 + z^2)^{-3/2}(2y) = -y(x^2 + y^2 + z^2)^{-3/2}$

$\frac{\partial f}{\partial z} = -\frac{1}{2}(x^2 + y^2 + z^2)^{-3/2}(2z) = -z(x^2 + y^2 + z^2)^{-3/2}. \ \frac{\partial^2 f}{\partial x^2} = -(x^2 + y^2 + z^2)^{-3/2} + 3x^2(x^2 + y^2 + z^2)^{-5/2}$

$\frac{\partial^2 f}{\partial y^2} = -(x^2 + y^2 + z^2)^{-3/2} + 3y^2(x^2 + y^2 + z^2)^{-5/2}, \frac{\partial^2 f}{\partial z^2} = -(x^2 + y^2 + z^2)^{-3/2} + 3z^2(x^2 + y^2 + z^2)^{-5/2}$

$\therefore \ \frac{\partial^2 f}{\partial x^2} + \frac{\partial^2 f}{\partial y^2} + \frac{\partial^2 f}{\partial z^2} = \left(-(x^2 + y^2 + z^2)^{-3/2} + 3x^2(x^2 + y^2 + z^2)^{-5/2}\right) +$

$\left(-(x^2 + y^2 + z^2)^{-3/2} + 3y^2(x^2 + y^2 + z^2)^{-5/2}\right) + \left(-(x^2 + y^2 + z^2)^{-3/2} + 3z^2(x^2 + y^2 + z^2)^{-5/2}\right) =$

$-3(x^2 + y^2 + z^2)^{-3/2} + (3x^2 + 3y^2 + 3z^2)(x^2 + y^2 + z^2)^{-5/2} = 0$

53. $\frac{\partial w}{\partial x} = \cos(x + ct) - 2\sin(2x + 2ct), \frac{\partial w}{\partial t} = c\cos(x + ct) - 2c\sin(2x + 2ct). \ \frac{\partial^2 w}{\partial x^2} = -\sin(x + ct) -$

$4\cos(2x + 2ct), \ \frac{\partial^2 w}{\partial t^2} = -c^2\sin(x + ct) - 4c^2\cos(2x + 2ct) \ \therefore \ \frac{\partial^2 w}{\partial t^2} = c^2(-\sin(x + ct) - 4\cos(2x + 2ct)) =$

$c^2\frac{\partial^2 w}{\partial x^2}$

13.4 DIFFERENTIABILITY, LINEARIZATION, AND DIFFERENTIALS

1. a) $f(0,0) = 1, f_x(x,y) = 2x \Rightarrow f_x(0,0) = 0, f_y(x,y) = 2y \Rightarrow f_y(0,0) = 0 \Rightarrow L(x,y) = 1 + 0(x - 0) + 0(y - 0) = 1$

   b) $f(1,1) = 3, f_x(1,1) = 2, f_y(1,1) = 2 \Rightarrow L(x,y) = 3 + 2(x - 1) + 2(y - 1) = 2x + 2y - 1$

5. a) $f(0,0) = 5, f_x(x,y) = 3$ for all (x,y), $f_y(x,y) = -4$ for all $(x,y) \Rightarrow L(x,y) = 5 + 3(x - 0) - 4(y - 0) =$

   $5 + 3x - 4y$

   b) $f(1,1) = 4, f_x(1,1) = 3, f_y(1,1) = -4 \Rightarrow L(x,y) = 4 + 3(x - 1) - 4(y - 1) = 3x - 4y + 5$

9. $f(0,0) = 1, f_x(x,y) = \cos y \Rightarrow f_x(0,0) = 1, f_y(x,y) = 1 - x\sin y \Rightarrow f_y(0,0) = 1 \Rightarrow L(x,y) = 1 + 1(x - 0) +$

   $1(y - 0) = x + y + 1. \ f_{xx}(x,y) = 0, f_{yy}(x,y) = -x\cos y, f_{xy}(x,y) = -\sin y \Rightarrow M = 1.$

   $\therefore \ |E(x,y)| \le \frac{1}{2}(1)(|x| + |y|)^2 \le \frac{1}{2}(0.2 + 0.2)^2 = 0.08$

13. Let the width, w, be the long side. Then $A = lw \Rightarrow dA = A_l \, dl + A_w \, dw \Rightarrow dA = w\,dl + l\,dw$. Since $w > l$, dA is more sensitive to a change in w than l. $\therefore$ pay more attention to the width.

17. $V_r = 2\pi rh, V_h = \pi r^2 \Rightarrow dV = V_r\,dr + V_h\,dh \Rightarrow dV = 2\pi rh\,dr + \pi r^2\,dh \Rightarrow dV|_{(5,12)} = 120\pi\,dr + 25\pi\,dh.$

    Since $|dr| \le 0.1$ cm, $|dh| \le 0.1$ cm, $dV \le 120\pi(0.1) + 25\pi(0.1) = 14.5\pi$ cm^3. $V(5,12) = 300\pi$ cm$^3 \Rightarrow$

    Maximum percentage error $= \pm\frac{14.5\pi}{300\pi} \text{ X } 100 = \pm 4.83\%$

21. $dR = \left(\frac{R}{R_1}\right)^2 dR_1 + \left(\frac{R}{R_2}\right)^2 dR_2$ (See Exercise 20 above). R_1 changes from 20 to 20.1 ohms $\Rightarrow dR_1 =$

    0.1 ohms, R_2 changes from 25 to 24.9 ohms $\Rightarrow dR_2 = -0.1$ ohms. $\frac{1}{R} = \frac{1}{R_1} + \frac{1}{R_2} \Rightarrow R = \frac{100}{9}$ ohms.

    $dR|_{(20,25)} = \frac{(100/9)^2}{(20)^2}(0.1) + \frac{(100/9)^2}{(25)^2}(-0.1) = 0.011$ ohms $\Rightarrow$ Percentage change $= \frac{dR}{R}\Big|_{(20,25)} \text{ X } 100$

    $= \frac{0.011}{100/9} \text{ X } 100 \approx 0.099\%$ or about 0.1%.

25. a) $f(1,0,0) = 1$, $f_x(1,0,0) = \frac{x}{\sqrt{x^2 + y^2 + z^2}}\Big|_{(1,0,0)} = 1$, $f_y(1,0,0) = \frac{y}{\sqrt{x^2 + y^2 + z^2}}\Big|_{(1,0,0)} = 0$,
$f_z(1,0,0) = \frac{z}{\sqrt{x^2 + y^2 + z^2}}\Big|_{(1,0,0)} = 0 \Rightarrow L(x,y,z) = x$

b) $f(1,1,0) = \sqrt{2}$, $f_x(1,1,0) = \frac{1}{\sqrt{2}}$, $f_y(1,1,0) = \frac{1}{\sqrt{2}}$, $f_z(1,1,0) = 0 \Rightarrow L(x,y,z) = \frac{1}{\sqrt{2}}x + \frac{1}{\sqrt{2}}y$

c) $f(1,2,2) = 3$, $f_x(1,2,2) = \frac{1}{3}$, $f_y(1,2,2) = \frac{2}{3}$, $f_z(1,2,2) = \frac{2}{3} \Rightarrow L(x,y,z) = \frac{1}{3}x + \frac{2}{3}y + \frac{2}{3}z$

29. $f(a,b,c,d) = \begin{vmatrix} a & b \\ c & d \end{vmatrix} = ad - bc \Rightarrow f_a = d$, $f_b = -c$, $f_c = -b$, $f_d = a \Rightarrow df = d\,da - c\,db - b\,dc + a\,dd$.

Since $|a|$ is much greater than $|b|$, $|c|$, and $|d|$, f is most sensitive to a change in d.

33. $u_x = e^y$, $u_y = xe^y + \sin z$, $u_z = y\cos z \Rightarrow du = e^y\,dx + (xe^y + \sin z)dy + (y\cos z)dz \Rightarrow$
$du\Big|_{(2,\ln 3,\pi/2)} = 3\,dx + 7\,dy + 0\,dz = 3\,dx + 7\,dy \Rightarrow$ magnitude of the maximum possible error $\le$
$3(0.2) + 7(0.6) = 4.8$

13.5 THE CHAIN RULE

1. a) $\frac{\partial w}{\partial x} = 2x$, $\frac{\partial w}{\partial y} = 2y$, $\frac{dx}{dt} = -\sin t$, $\frac{dy}{dt} = \cos t \Rightarrow \frac{dw}{dt} = -2x\sin t + 2y\cos t = -2\cos t\sin t + 2\sin t\cos t = 0$
$w = x^2 + y^2 = \cos^2 t + \sin^2 t = 1 \Rightarrow \frac{dw}{dt} = 0$

b) $\frac{dw}{dt}(\pi) = 0$

5. a) $\frac{\partial w}{\partial x} = 2ye^x$, $\frac{\partial w}{\partial y} = 2e^x$, $\frac{\partial w}{\partial z} = -\frac{1}{z}$, $\frac{dx}{dt} = \frac{2t}{t^2+1}$, $\frac{dy}{dt} = \frac{1}{t^2+1}$, $\frac{dz}{dt} = e^t \Rightarrow \frac{dw}{dt} = \frac{4yte^x}{t^2+1} + \frac{2e^x}{t^2+1} - \frac{e^t}{z} =$
$\frac{4t\tan^{-1}t\, e^{\ln(t^2+1)}}{t^2+1} + \frac{2(t^2+1)}{t^2+1} - \frac{e^t}{e^t} = 4t\tan^{-1}t + 1$. $w = 2ye^x - \ln z = 2\tan^{-1}t\, e^{\ln(t^2+1)} - \ln e^t =$
$(2\tan^{-1}t)(t^2+1) - t \Rightarrow \frac{dw}{dt} = \left(\frac{2}{t^2+1}\right)\left(t^2+1\right) + 2t\left(2\tan^{-1}t\right)(2t) - 1 = 4t\tan^{-1}t + 1$

b) $\frac{dw}{dt}(1) = \pi + 1$

9. a) $w = xy + yz + xz$, $x = u + v$, $y = u - v$, $z = uv \Rightarrow \frac{\partial w}{\partial u} = (y+z)(1) + (x+z)(1) + (y+x)(v) = x + y + 2z +$
$v(y+x)$. As a function of u an v only, $\frac{\partial w}{\partial u} = u + v + u - v + 2uv + v(u - v + u + v) = 2u + 4uv$.
$\frac{\partial w}{\partial v} = (y+z)(1) + (x+z)(-1) + (y+x)(u) = y - x + (y+x)u$. As a function of u and v only, $\frac{\partial w}{\partial v} = u - v -$
$(u+v) + (u - v + u + v)u = -2v + 2u^2$. To find the partial derivatives directly, let $w = (u+v)(u-v) +$
$(u-v)uv + (u+v)uv = u^2 - v^2 + 2u^2v$. Then find the partial with respect to u and the partial with respect
to v. The answers will be the same as above.

b) At $\left(\frac{1}{2}, 1\right)$, $\frac{\partial w}{\partial u} = 2\left(\frac{1}{2}\right) + 4\left(\frac{1}{2}\right)(1) = 3$, $\frac{\partial w}{\partial v} = -2(1) + 2\left(\frac{1}{2}\right)^2 = -\frac{3}{2}$.

13. $\frac{dz}{dt} = \frac{\partial z}{\partial x}\frac{dx}{dt} + \frac{\partial z}{\partial y}\frac{dy}{dt}$

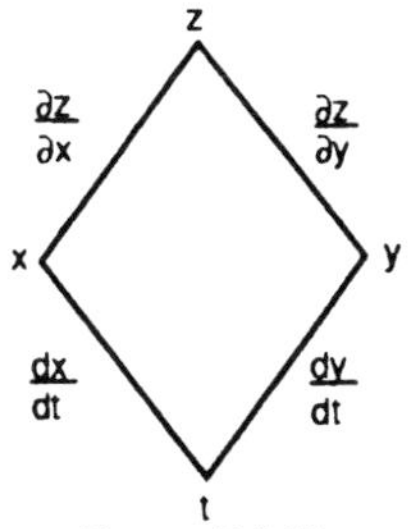

Diagram 13.5.13

17. $\frac{\partial w}{\partial u} = \frac{\partial w}{\partial x}\frac{\partial x}{\partial u} + \frac{\partial w}{\partial y}\frac{\partial y}{\partial u}$

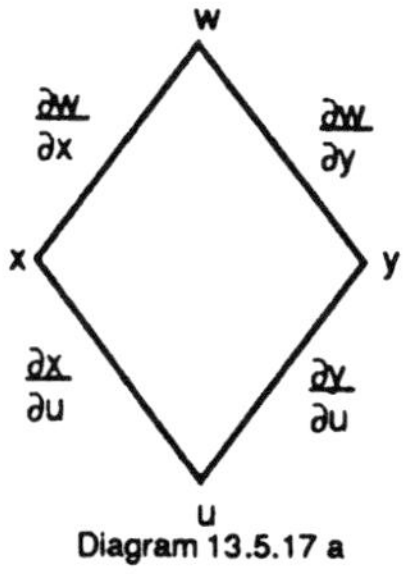

Diagram 13.5.17 a

$\frac{\partial w}{\partial v} = \frac{\partial w}{\partial x}\frac{\partial x}{\partial v} + \frac{\partial w}{\partial y}\frac{\partial y}{\partial v}$

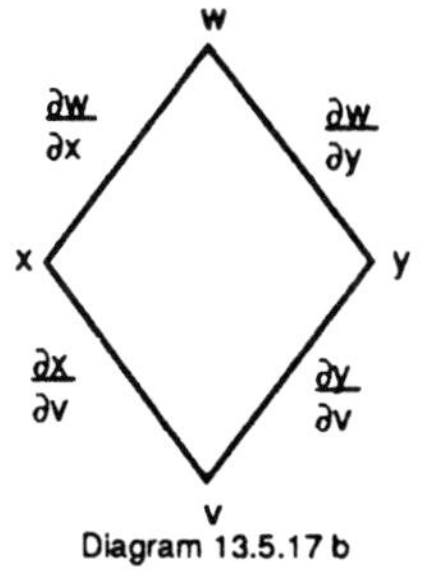

Diagram 13.5.17 b

21. $\frac{\partial w}{\partial s} = \frac{dw}{du}\frac{\partial u}{\partial s}$

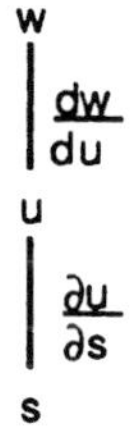

Diagram 13.5.21 a

$\frac{\partial w}{\partial t} = \frac{dw}{du}\frac{\partial u}{\partial t}$

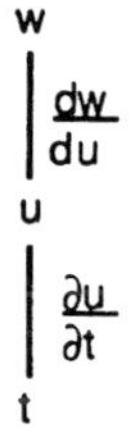

Diagram 13.5.21 b

25. Let $F(x,y) = x^3 - 2y^2 + xy = 0 \Rightarrow F_x(x,y) = 3x^2 + y$, $F_y(x,y) = -4y + x \Rightarrow \frac{dy}{dx} = -\frac{F_x}{F_y} = -\frac{3x^2 + y}{(-4y + x)} \Rightarrow$
$\frac{dy}{dx}(1,1) = \frac{4}{3}$

29. Let $F(x,y,z)) = z^3 - xy + yz + y^3 - 2 = 0 \Rightarrow F_x(x,y,z) = -y$, $F_y(x,y,z) = -x + z + 3y^2$, $F_z(x,y,z) = 3z^2 + y \Rightarrow$
$\frac{\partial z}{\partial x} = -\frac{F_x}{F_z} = -\frac{-y}{3z^2 + y} = \frac{y}{3z^2 + y} \Rightarrow \frac{\partial z}{\partial x}(1,1,1) = \frac{1}{4}$. $\frac{\partial z}{\partial y} = -\frac{F_y}{F_z} = -\frac{-x + z + 3y^2}{3z^2 + y} = \frac{x - z - 3y^2}{3z^2 + y} \Rightarrow$
$\frac{\partial z}{\partial y}(1,1,1) = -\frac{3}{4}$

33. $\frac{\partial w}{\partial r} = \frac{\partial w}{\partial x}\frac{\partial x}{\partial r} + \frac{\partial w}{\partial y}\frac{\partial y}{\partial r} + \frac{\partial w}{\partial z}\frac{\partial z}{\partial r} = 2(x + y + z)(1) + 2(x + y + z)(-\sin(r + s)) + 2(x + y + z)(\cos(r + s)) =$
$2(x + y + z)(1 - \sin(r + s) + \cos(r + s)) = 2(r - s + \cos(r + s) + \sin(r + s))(1 - \sin(r + s) + \cos(r + s)) \Rightarrow$
$\left.\frac{\partial w}{\partial r}\right|_{r=1, s=-1} = 12$

37. $\frac{\partial z}{\partial u} = \frac{dz}{dx}\frac{\partial x}{\partial u} = \frac{5}{1 + x^2}e^u = \frac{5}{1 + \left(e^u + \ln v\right)^2}e^u \Rightarrow \left.\frac{\partial z}{\partial u}\right|_{u=\ln 2, v=1} = 2$
$\frac{\partial z}{\partial v} = \frac{dz}{dx}\frac{\partial x}{\partial v} = \frac{5}{1 + x^2}\left(\frac{1}{v}\right) = \frac{5}{1 + \left(e^u + \ln v\right)^2}\left(\frac{1}{v}\right) \Rightarrow \left.\frac{\partial z}{\partial v}\right|_{u=\ln 2, v=1} = 1$

41. $f_x(x,y,z) = \cos t$, $f_y(x,y,z) = \sin t$, $f_z(x,y,z) = t^2 + t - 2$. $\frac{df}{dt} = \frac{\partial f}{\partial x}\frac{dx}{dt} + \frac{\partial f}{\partial y}\frac{dy}{dt} + \frac{\partial f}{\partial z}\frac{dz}{dt} = (\cos t)(-\sin t) +$
$(\sin t)(\cos t) + (t^2 + t - 2)(1) = t^2 + t - 2$. $\frac{df}{dt} = 0 \Rightarrow t^2 + t - 2 = 0 \Rightarrow t = -2$ or $t = 1$
$t = -2 \Rightarrow x = \cos(-2)$, $y = \sin(-2)$, $z = -2$; $t = 1 \Rightarrow x = \cos 1$, $y = \sin 1$, $z = 1$

13.6 DIRECTIONAL DERIVATIVES, GRADIENT VECTORS, AND TANGENT PLANES

1. $\frac{\partial f}{\partial x} = -1, \frac{\partial f}{\partial y} = 1 \Rightarrow \nabla f = -\mathbf{i} + \mathbf{j}$

$-1 = y - x$ is the level curve.

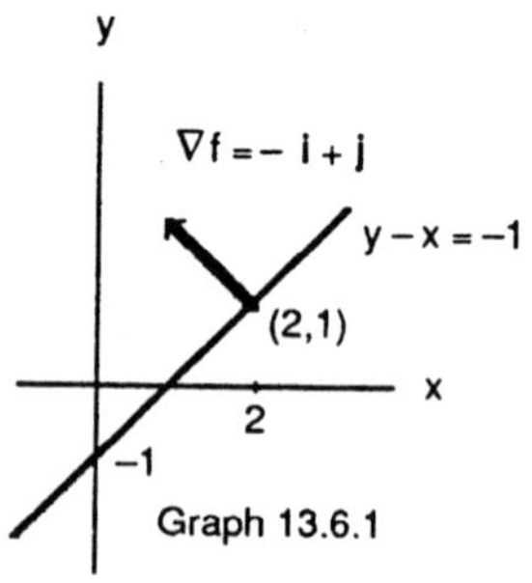

Graph 13.6.1

5. $\frac{\partial f}{\partial x} = 2x \Rightarrow \frac{\partial f}{\partial x}(1,1,1) = 2.\ \frac{\partial f}{\partial y} = 2y \Rightarrow \frac{\partial f}{\partial y}(1,1,1) = 2.\ \frac{\partial f}{\partial z} = -4z \Rightarrow \frac{\partial f}{\partial z}(1,1,1) = -4 \Rightarrow \nabla f = 2\mathbf{i} + 2\mathbf{j} - 4\mathbf{k}$

9. $\mathbf{u} = \frac{\mathbf{A}}{|\mathbf{A}|} = \frac{12\mathbf{i} + 5\mathbf{j}}{\sqrt{12^2 + 5^2}} = \frac{12}{13}\mathbf{i} + \frac{5}{13}\mathbf{j}.\ f_x(x,y) = 1 + \frac{y^2}{x^2} \Rightarrow f_x(1,1) = 2, f_y(x,y) = -\frac{2y}{x} \Rightarrow f_y(1,1) = -2 \Rightarrow$
$\nabla f = 2\mathbf{i} - 2\mathbf{j} \Rightarrow (D_\mathbf{u}f)_{P_0} = \nabla f \cdot \mathbf{u} = \frac{24}{13} - \frac{10}{13} = \frac{14}{13}$

13. $\mathbf{u} = \frac{\mathbf{A}}{|\mathbf{A}|} = \frac{3\mathbf{i} + 6\mathbf{j} - 2\mathbf{k}}{\sqrt{3^2 + 6^2 + (-2)^2}} = \frac{3}{7}\mathbf{i} + \frac{6}{7}\mathbf{j} - \frac{2}{7}\mathbf{k}.\ f_x(x,y,z) = y + z \Rightarrow f_x(1,-1,2) = 1, f_y(x,y,z) = x + z \Rightarrow$
$f_y(1,-1,2) = 3, f_z(x,y,z) = y + x \Rightarrow f_z(1,-1,2) = 0 \Rightarrow \nabla f = \mathbf{i} + 3\mathbf{j} \Rightarrow (D_\mathbf{u}f)_{P_0} = \nabla f \cdot \mathbf{u} = \frac{3}{7} + \frac{18}{7} = 3$

17. $\nabla f = (2x + y)\mathbf{i} + (x + 2y)\mathbf{j} \Rightarrow \nabla f(-1,1) = -\mathbf{i} + \mathbf{j} \Rightarrow \mathbf{u} = \frac{\nabla f}{|\nabla f|} = \frac{-\mathbf{i} + \mathbf{j}}{\sqrt{(-1)^2 + 1^2}} = -\frac{1}{\sqrt{2}}\mathbf{i} + \frac{1}{\sqrt{2}}\mathbf{j}$. f increases most rapidly in the direction $\mathbf{u} = -\frac{1}{\sqrt{2}}\mathbf{i} + \frac{1}{\sqrt{2}}\mathbf{j}$ and decreases most rapidly in the direction $-\mathbf{u} = \frac{1}{\sqrt{2}}\mathbf{i} - \frac{1}{\sqrt{2}}\mathbf{j}$. $(D_\mathbf{u}f)_{P_0} = \nabla f \cdot \mathbf{u} = \sqrt{2}$, $(D_{-\mathbf{u}}f)_{P_0} = -\sqrt{2}$

21. $\nabla g = e^y\mathbf{i} + xe^y\mathbf{j} + 2z\mathbf{k} \Rightarrow \nabla g\left(1, \ln 2, \frac{1}{2}\right) = 2\mathbf{i} + 2\mathbf{j} + \mathbf{k} \Rightarrow \mathbf{u} = \frac{\nabla g}{|\nabla g|} = \frac{2\mathbf{i} + 2\mathbf{j} + \mathbf{k}}{\sqrt{2^2 + 2^2 + 1^2}} = \frac{2}{3}\mathbf{i} + \frac{2}{3}\mathbf{j} + \frac{1}{3}\mathbf{k}.$

g increases most rapidly in the direction $\mathbf{u} = \frac{2}{3}\mathbf{i} + \frac{2}{3}\mathbf{j} + \frac{1}{3}\mathbf{k}$; decreases most rapidly in the direction $-\mathbf{u} = -\frac{2}{3}\mathbf{i} - \frac{2}{3}\mathbf{j} - \frac{1}{3}\mathbf{k}$. $(D_\mathbf{u}g)_{P_0} = \nabla g \cdot \mathbf{u} = 3$; $(D_{-\mathbf{u}}g)_{P_0} = -3$

25. $\mathbf{A} = \overrightarrow{P_0P_1} = -2\mathbf{i} + 2\mathbf{j} + 2\mathbf{k}$. $\nabla g = (1 + \cos z)\mathbf{i} + (1 - \sin z)\mathbf{j} + (-x \sin z - y \cos z)\mathbf{k} \Rightarrow \nabla g(2,-1,0) = 2\mathbf{i} + \mathbf{j} + \mathbf{k}$. $\mathbf{u} = \frac{\mathbf{A}}{|\mathbf{A}|} = \frac{-2\mathbf{i} + 2\mathbf{j} + 2\mathbf{k}}{\sqrt{(-2)^2 + 2^2 + 2^2}} = -\frac{1}{\sqrt{3}}\mathbf{i} + \frac{1}{\sqrt{3}}\mathbf{j} + \frac{1}{\sqrt{3}}\mathbf{k} \Rightarrow \nabla g \cdot \mathbf{u} = 0.\ \therefore\ df = (\nabla g \cdot \mathbf{u})ds = 0(0.2) = 0$

29. $\nabla f = -2x\mathbf{i} + 2\mathbf{k} \Rightarrow \nabla f(2,0,2) = -4\mathbf{i} + 2\mathbf{k} \Rightarrow$ Tangent plane: $-4(x - 2) + 2(z - 2) = 0 \Rightarrow -4x + 2z + 4 = 0 \Rightarrow$
$-2x + z + 2 = 0$; Normal line: $x = 2 - 4t, y = 0, z = 2 + 2t$

33. $\nabla f = \mathbf{i} + \mathbf{j} + \mathbf{k}$ for all points $\Rightarrow$ Tangent plane: $1(x - 0) + 1(y - 1) + 1(z - 0) = 0 \Rightarrow x + y + z - 1 = 0$;
Normal line: $x = t,\ y = 1 + t,\ z = t$

37. $z = f(x,y) = \sqrt{y - x} \Rightarrow f_x(x,y) = -\frac{1}{2}(y-x)^{-1/2},\ f_y(x,y) = \frac{1}{2}(y-x)^{-1/2} \Rightarrow f_x(1,2) = -\frac{1}{2},\ f_y(1,2) = \frac{1}{2} \Rightarrow$ Tangent Plane at $(1,2,1)$: $-\frac{1}{2}(x-1) + \frac{1}{2}(y-2) - (z-1) = 0 \Rightarrow -\frac{1}{2}x + \frac{1}{2}y - z + \frac{1}{2} = 0$ or $x - y + 2z - 1 = 0$.

41. $\nabla f = y\,\mathbf{i} + x\,\mathbf{j} \Rightarrow \nabla f(2,-2) = -2\,\mathbf{i} + 2\,\mathbf{j} \Rightarrow$ Tangent line: $-2(x - 2) + 2(y + 2) = 0 \Rightarrow y = x - 4$

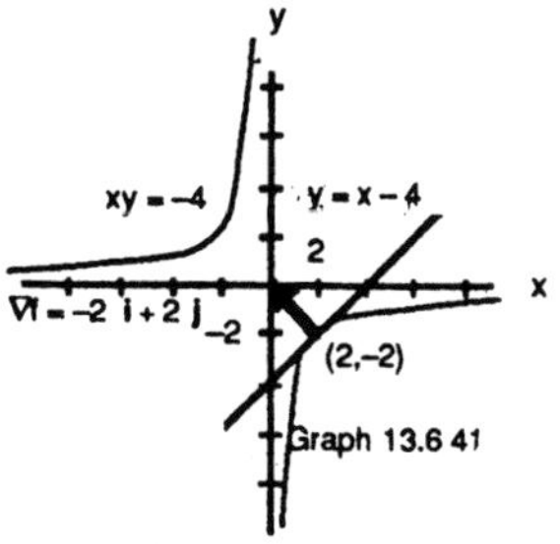

Graph 13.6 41

45. $\nabla f = 2x\,\mathbf{i} + 2\,\mathbf{j} + 2\,\mathbf{k} \Rightarrow \nabla f(1,1,\frac{1}{2}) = 2\,\mathbf{i} + 2\,\mathbf{j} + 2\,\mathbf{k}$. $\nabla g = \mathbf{j}$ for all points P. $\mathbf{v} = \nabla f \times \nabla g \Rightarrow$

$$\mathbf{v} = \begin{vmatrix} \mathbf{i} & \mathbf{j} & \mathbf{k} \\ 2 & 2 & 2 \\ 0 & 1 & 0 \end{vmatrix} = -2\,\mathbf{i} + 2\,\mathbf{k} \Rightarrow \text{Tangent line: } x = 1 - 2t,\ y = 1,\ z = \frac{1}{2} + 2t$$

49. $\nabla f = y\,\mathbf{i} + (x + 2y)\,\mathbf{j} \Rightarrow \nabla f(3,2) = 2\,\mathbf{i} + 7\,\mathbf{j}$. $\mathbf{A}$, orthogonal to ∇f, is $\mathbf{A} = 7\,\mathbf{i} - 2\,\mathbf{j} \Rightarrow \mathbf{u} = \frac{\mathbf{A}}{|\mathbf{A}|} = \frac{7\,\mathbf{i} - 2\,\mathbf{j}}{\sqrt{7^2 + (-2)^2}} = \frac{7}{\sqrt{53}}\mathbf{i} - \frac{2}{\sqrt{53}}\mathbf{j} \Rightarrow -\mathbf{u} = -\frac{7}{\sqrt{53}}\mathbf{i} + \frac{2}{\sqrt{53}}\mathbf{j}$

53. $\nabla f = f_x(1,2)\,\mathbf{i} + f_y(1,2)\,\mathbf{j}$. $\mathbf{u}_1 = \frac{\mathbf{i} + \mathbf{j}}{\sqrt{1^2 + 1^2}} = \frac{1}{\sqrt{2}}\mathbf{i} + \frac{1}{\sqrt{2}}\mathbf{j}$. $\left(D_{\mathbf{u}_1}f\right)(1,2) = f_x(1,2)\left(\frac{1}{\sqrt{2}}\right) + f_y(1,2)\left(\frac{1}{\sqrt{2}}\right) = 2\sqrt{2} \Rightarrow f_x(1,2) + f_y(1,2) = 4$. $\mathbf{u}_2 = -\mathbf{j}$. $\left(D_{\mathbf{u}_2}f\right)(1,2) = f_x(1,2)(0) + f_y(1,2)(-1) = -3 \Rightarrow -f_y(1,2) = -3 \Rightarrow f_y(1,2) = 3$. $\therefore$ $f_x(1,2) + 3 = 4 \Rightarrow f_x(1,2) = 1$. Then $\nabla f(1,2) = \mathbf{i} + 3\,\mathbf{j}$.

$$\mathbf{u} = \frac{\mathbf{A}}{|\mathbf{A}|} = \frac{-\mathbf{i} - 2\,\mathbf{j}}{\sqrt{(-1)^2 + (-2)^2}} = -\frac{1}{\sqrt{5}}\mathbf{i} - \frac{2}{\sqrt{5}}\mathbf{j} \Rightarrow \left(D_{\mathbf{u}}f\right)_{P_0} = \nabla f \cdot \mathbf{u} = -\frac{1}{\sqrt{5}} - \frac{6}{\sqrt{5}} = -\frac{7}{\sqrt{5}}$$

57. a) $\mathbf{r} = \sqrt{t}\,\mathbf{i} + \sqrt{t}\,\mathbf{j} - \frac{1}{4}(t + 3)\,\mathbf{k} \Rightarrow \mathbf{v} = \frac{1}{2}t^{-1/2}\,\mathbf{i} + \frac{1}{2}t^{-1/2}\,\mathbf{j} - \frac{1}{4}\mathbf{k}$. $t = 1 \Rightarrow x = 1, y = 1, z = -1 \Rightarrow P_0 = (1,1,-1)$. Also $t = 1 \Rightarrow \mathbf{v}(1) = \frac{1}{2}\mathbf{i} + \frac{1}{2}\mathbf{j} - \frac{1}{4}\mathbf{k}$. $f(x,y,z) = x^2 + y^2 - z - 3 = 0 \Rightarrow \nabla f = 2x\,\mathbf{i} + 2y\,\mathbf{j} - \mathbf{k} \Rightarrow \nabla f(1,1,-1) = 2\,\mathbf{i} + 2\,\mathbf{j} - \mathbf{k}$. $\therefore$ $\mathbf{v} = \frac{1}{4}(\nabla f) \Rightarrow$ The curve is normal to the surface.

b) $\mathbf{r} = \sqrt{t}\,\mathbf{i} + \sqrt{t}\,\mathbf{j} + (2t - 1)\,\mathbf{k} \Rightarrow \mathbf{v} = \frac{1}{2}t^{-1/2}\,\mathbf{i} + \frac{1}{2}t^{-1/2}\,\mathbf{j} + 2\,\mathbf{k}$. $t = 1 \Rightarrow x = 1, y = 1, z = 1$. Also $t = 1 \Rightarrow \mathbf{v}(1) = \frac{1}{2}\mathbf{i} + \frac{1}{2}\mathbf{j} + 2\,\mathbf{k}$. $f(x,y,z) = x^2 + y^2 - z - 1 = 0 \Rightarrow \nabla f = 2x\,\mathbf{i} + 2y\,\mathbf{j} - \mathbf{k} \Rightarrow \nabla f(1,1,1) = 2\,\mathbf{i} + 2\,\mathbf{j} - \mathbf{k} \Rightarrow \mathbf{v} \cdot \nabla f = \frac{1}{2}(2) + \frac{1}{2}(2) + 2(-1) = 0 \Rightarrow$ The curve is tangent to the surface when $t = 1$.

13.7 MAXIMA, MINIMA, AND SADDLE POINTS

1. $f_x(x,y) = 2x + y + 3 = 0$ and $f_y(x,y) = x + 2y - 3 = 0 \Rightarrow x = -3, y = 3 \Rightarrow$ critical point is $(-3,3)$. $f_{xx}(-3,3) = 2$, $f_{yy}(-3,3) = 2$, $f_{xy}(-3,3) = 1 \Rightarrow f_{xx}f_{yy} - f_{xy}^2 = 3 > 0$ and $f_{xx} > 0 \Rightarrow$ local minimum. $f(-3,3) = -5$

5. $f_x(x,y) = 2x + y + 3 = 0$ and $f_y(x,y) = x + 2 = 0 \Rightarrow x = -2, y = 1 \Rightarrow$ critical point is $(-2,1)$. $f_{xx}(-2,1) = 2$, $f_{yy}(-2,1) = 0$, $f_{xy}(-2,1) = 1 \Rightarrow f_{xx}f_{yy} - f_{xy}^2 = -1 \Rightarrow$ saddle point.

9. $f_x(x,y) = 2x - 4y = 0$ and $f_y(x,y) = -4x + 2y + 6 = 0 \Rightarrow x = 2, y = 1 \Rightarrow$ critical point is $(2,1)$. $f_{xx}(2,1) = 2$, $f_{yy}(2,1) = 2$, $f_{xy}(2,1) = -4 \Rightarrow f_{xx}f_{yy} - f_{xy}^2 = -12 \Rightarrow$ saddle point.

13. $f_x(x,y) = 2x - 2 = 0$ and $f_y(x,y) = -2y + 4 = 0 \Rightarrow x = 1, y = 2 \Rightarrow$ critical point is $(1,2)$. $f_{xx}(1,2) = 2$, $f_{yy}(1,2) = -2$, $f_{xy}(1,2) = 0 \Rightarrow f_{xx}f_{yy} - f_{xy}^2 = -4 \Rightarrow$ saddle point.

17. $f_x(x,y) = 3x^2 - 2y = 0$ and $f_y(x,y) = -3y^2 - 2x = 0 \Rightarrow x = 0, y = 0$ or $x = -\frac{2}{3}, y = \frac{2}{3} \Rightarrow$ critical points are $(0,0)$ and $\left(-\frac{2}{3}, \frac{2}{3}\right)$. For $(0,0)$: $f_{xx}(0,0) = 6x\big|_{(0,0)} = 0$, $f_{yy}(0,0) = -6y\big|_{(0,0)} = 0$, $f_{xy}(0,0) = -2 \Rightarrow f_{xx}f_{yy} - f_{xy}^2 = -4 \Rightarrow$ saddle point. For $\left(-\frac{2}{3}, \frac{2}{3}\right)$: $f_{xx}\left(-\frac{2}{3}, \frac{2}{3}\right) = -4$, $f_{yy}\left(-\frac{2}{3}, \frac{2}{3}\right) = -4$, $f_{xy}\left(-\frac{2}{3}, \frac{2}{3}\right) = -2$ $\Rightarrow f_{xx}f_{yy} - f_{xy}^2 = 12 > 0$ and $f_{xx} < 0 \Rightarrow$ local maximum. $f\left(-\frac{2}{3}, \frac{2}{3}\right) = \frac{170}{27}$

21. $f_x(x,y) = 27x^2 - 4y = 0$ and $f_y(x,y) = y^2 - 4x = 0 \Rightarrow x = 0, y = 0$ or $x = \frac{4}{9}, y = \frac{4}{3} \Rightarrow$ critical points are $(0,0)$ and $\left(\frac{4}{9}, \frac{4}{3}\right)$. For $(0,0)$: $f_{xx}(0,0) = 54x\big|_{(0,0)} = 0$, $f_{yy}(0,0) = 2y\big|_{(0,0)} = 0$, $f_{xy}(0,0) = -4 \Rightarrow f_{xx}f_{yy} - f_{xy}^2 = -16 \Rightarrow$ saddle point. For $\left(\frac{4}{9}, \frac{4}{3}\right)$: $f_{xx}\left(\frac{4}{9}, \frac{4}{3}\right) = 24$, $f_{yy}\left(\frac{4}{9}, \frac{4}{3}\right) = \frac{8}{3}$, $f_{xy}\left(\frac{4}{9}, \frac{4}{3}\right) = -4 \Rightarrow$ $f_{xx}f_{yy} - f_{xy}^2 = 48 > 0$ and $f_{xx} > 0 \Rightarrow$ local minimum. $f\left(\frac{4}{9}, \frac{4}{3}\right) = -\frac{64}{81}$

25. $f_x(x,y) = 4y - 4x^3 = 0$ and $f_y(x,y) = 4x - 4y^3 = 0 \Rightarrow x = 0, y = 0$ or $x = 1, y = 1$ or $x = -1, y = -1 \Rightarrow$ critical points are $(0,0)$, $(1,1)$, and $(-1,-1)$. For $(0,0)$: $f_{xx}(0,0) = -12x^2\big|_{(0,0)} = 0$, $f_{yy}(0,0) = -12y^2\big|_{(0,0)} = 0$, $f_{xy}(0,0) = 4 \Rightarrow$ $f_{xx}f_{yy} - f_{xy}^2 = -16 \Rightarrow$ saddle point. For $(1,1)$: $f_{xx}(1,1) = -12$, $f_{yy}(1,1) = -12$, $f_{xy}(1,1) = 4 \Rightarrow$ $f_{xx}f_{yy} - f_{xy}^2 = 128 > 0$ and $f_{xx} < 0 \Rightarrow$ local maximum. $f(1,1) = 2$. For $(-1,-1)$: $f_{xx}(-1,-1) = -12$, $f_{yy}(-1,-1) = -12$, $f_{xy}(-1,-1) = 4 \Rightarrow f_{xx}f_{yy} - f_{xy}^2 = 128 > 0$ and $f_{xx} < 0 \Rightarrow$ local maximum. $f(-1,-1) = 2$

29.

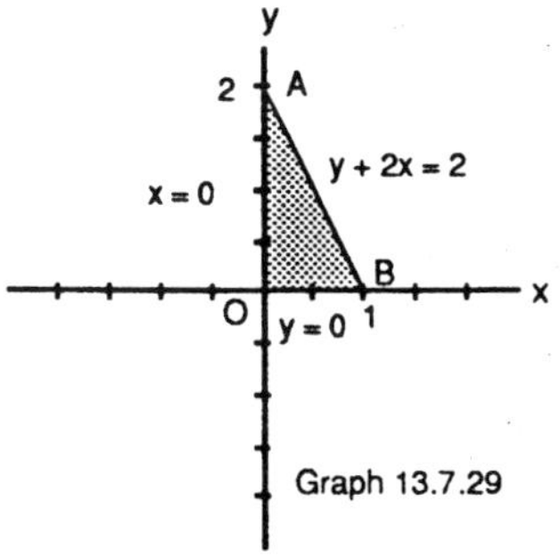

Graph 13.7.29

1. On OA, $f(x,y) = y^2 = f(0,y)$ on $0 \le y \le 2$. $f(0,0) = 0$. $f(0,2) = 4$. $f'(0,y) = 2y = 0 \Rightarrow y = 0,\ x = 0 \Rightarrow (0,0)$.
2. On OB, $f(x,y) = x^2 = f(x,0)$ on $0 \le x \le 1$. $f(1,0) = 1$. $f'(x,0) = 2x = 0 \Rightarrow x = 0,\ y = 0 \Rightarrow (0,0)$
3. On AB, $f(x,y) = 5x^2 - 8x + 4 = f(x,-2x+2)$ on $0 \le x \le 1$. $f(0,2) = 4$. $f'(x,-2x+2) = 10x - 8 = 0 \Rightarrow x = \frac{4}{5},\ y = \frac{2}{5}$. $f\left(\frac{4}{5}, \frac{2}{5}\right) = \frac{4}{5}$
4. For interior points of the triangular region, $f_x(x,y) = 2x = 0$ and $f_y(x,y) = 2y = 0 \Rightarrow x = 0,\ y = 0 \Rightarrow (0,0)$, not an interior point of the region. $\therefore$ absolute maximum is 4 at (0,2); absolute minimum is 0 at (0,0)

33.

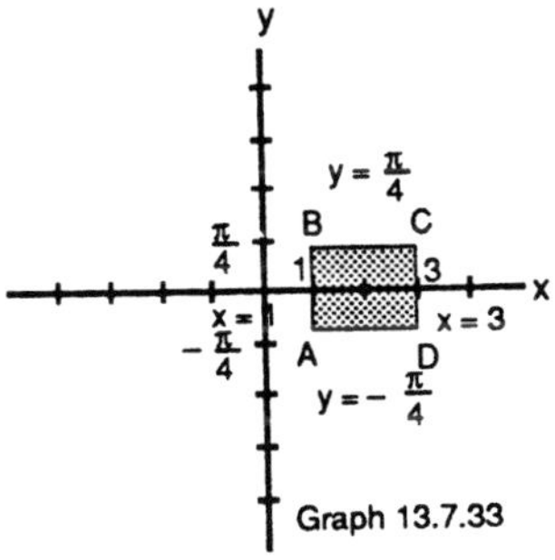

Graph 13.7.33

1. On AB, $f(x,y) = 3\cos y = f(1,y)$ on $-\frac{\pi}{4} \le y \le \frac{\pi}{4}$. $f\left(1, -\frac{\pi}{4}\right) = \frac{3\sqrt{2}}{2}$. $f\left(1, \frac{\pi}{4}\right) = \frac{3\sqrt{2}}{2}$. $f'(1,y) = -3\sin y = 0 \Rightarrow y = 0,\ x = 1 \Rightarrow f(1,0) = 3$.
2. On CD, $f(x,y) = 3\cos y = f(3,y)$ on $-\frac{\pi}{4} \le y \le \frac{\pi}{4}$. $f\left(3, -\frac{\pi}{4}\right) = \frac{3\sqrt{2}}{2}$. $f\left(3, \frac{\pi}{4}\right) = \frac{3\sqrt{2}}{2}$. $f'(3,y) = -3\sin y = 0 \Rightarrow y = 0,\ x = 3$. $f(3,0) = 3$.
3. On BC, $f(x,y) = \frac{\sqrt{2}}{2}(4x - x^2) = f\left(x, \frac{\pi}{4}\right)$ on $1 \le x \le 3$. $f'\left(x, \frac{\pi}{4}\right) = \sqrt{2}(2 - x) = 0 \Rightarrow x = 2,\ y = \frac{\pi}{4}$. $f\left(2, \frac{\pi}{4}\right) = 2\sqrt{2}$.
4. On AD, $f(x,y) = \frac{\sqrt{2}}{2}(4x - x^2) = f\left(x, -\frac{\pi}{4}\right)$ on $1 \le x \le 3$. $f'\left(x, -\frac{\pi}{4}\right) = \sqrt{2}(2 - x) = 0 \Rightarrow x = 2,\ y = -\frac{\pi}{4}$. $f\left(2, -\frac{\pi}{4}\right) = 2\sqrt{2}$.
5. For interior points of the region, $f_x(x,y) = (4 - 2x)\cos y = 0$ and $f_y(x,y) = -(4x - x^2)\sin y = 0 \Rightarrow x = 2$, $y = 0 \Rightarrow f(2,0) = 4$.
   $\therefore$ the absolute maximum is 4 at (2,0) and the absolute minimum is $\frac{3\sqrt{2}}{2}$ at $\left(3, -\frac{\pi}{4}\right)$, $\left(3, \frac{\pi}{4}\right)$, $\left(1, -\frac{\pi}{4}\right)$, and $\left(1, \frac{\pi}{4}\right)$.

37. a) $f_x(x,y) = 2x - 4y = 0$ and $f_y(x,y) = 2y - 4x = 0 \Rightarrow x = 0,\ y = 0$. $f_{xx}(0,0) = 2$, $f_{yy}(0,0) = 2$, $f_{xy}(0,0) = -4$ $\Rightarrow f_{xx}f_{yy} - f_{xy}^2 = -12 \Rightarrow$ saddle point.

b) $f_x(x,y) = 2x - 2 = 0$ and $f_y(x,y) = 2y - 4 = 0 \Rightarrow x = 1,\ y = 2$. $f_{xx}(1,2) = 2$, $f_{yy}(1,2) = 2$, $f_{xy}(1,2) = 0 \Rightarrow$ $f_{xx}f_{yy} - f_{xy}^2 = 4 > 0$ and $f_{xx} > 0 \Rightarrow$ local minimum at (1,2).

c) $f_x(x,y) = 9x^2 - 9 = 0$ and $f_y(x,y) = 2y + 4 = 0 \Rightarrow x = \pm 1,\ y = -2$. For (1,−2), $f_{xx}(1,-2) = 18x\big|_{(1,-2)} = 18$, $f_{yy}(1,-2) = 2$, $f_{xy}(1,-2) = 0 \Rightarrow f_{xx}f_{yy} - f_{xy}^2 = 36 > 0$ and $f_{xx} > 0 \Rightarrow$ local minimum at (1,−2).
For (−1,−2), $f_{xx}(-1,-2) = -18$, $f_{yy}(-1,-2) = 2$, $f_{xy}(-1,-2) = 0 \Rightarrow f_{xx}f_{yy} - f_{xy}^2 = -36 \Rightarrow$ saddle point.

41. No: for example $f(x,y) = xy$ has a saddle point at $(a,b) = (0,0)$ where $f_x = f_y = 0$.

45. a) $x = 2t,\ y = t + 1 \Rightarrow f(t) = 2t^2 + 2t \Rightarrow \frac{df}{dt} = 4t + 2.\ \frac{df}{dt} = 0 \Rightarrow t = -\frac{1}{2}.\ f''(t) = 4 \Rightarrow f\left(-\frac{1}{2}\right)$ is a minimum.

$f\left(-\frac{1}{2}\right) = -\frac{1}{2}$ is an absolute minimum since f(t) is an upward parabola. No absolute maximum.

b) $x = 2t,\ y = t + 1$ on $-1 \le t \le 0 \Rightarrow t = -\frac{1}{2}$ is a critical number on the interval (see part a) above).

$f\left(-\frac{1}{2}\right) = -\frac{1}{2}.\ f(0) = 0,\ f(-1) = 0 \Rightarrow$ absolute minimum is $-\frac{1}{2}$ at $t = -\frac{1}{2}$; absolute maximum is 0 at $t = 0, -1$.

c) $x = 2t,\ y = t + 1$ on $0 \le t \le 1 \Rightarrow$ no critical numbers on the interval (see part a) above). $f(0) = 0$,

$f(1) = 4 \Rightarrow$ absolute minimum is 0 at $t = 0$; absolute maximum is 4 at $t = 1$.

49.

k	x_k	y_k	x_k^2	$x_k y_k$
1	0	0	0	0
2	1	2	1	2
3	2	3	4	6
Σ	3	5	5	8

$m = \frac{3(5) - 3(8)}{3^2 - 3(5)} = 1.5,\ b = \frac{1}{3}(5 - 1.5(3)) = \frac{1}{6}$

$\therefore\ y = \frac{3}{2}x + \frac{1}{6}.\ y\big|_{x=4} = \frac{37}{6}$

53. a)

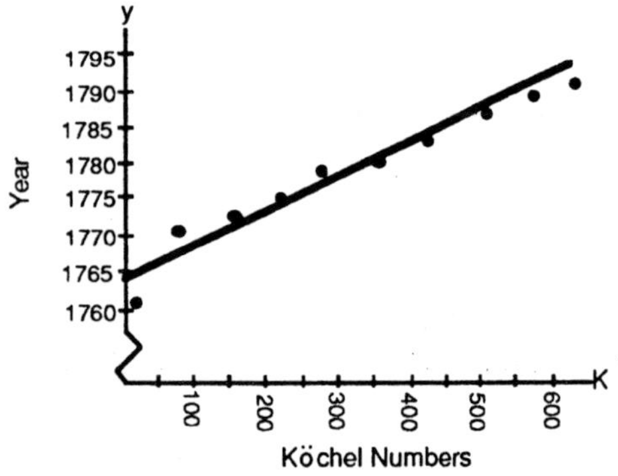

Graph 13.7.53 a

b) See the Table on the next page. $m = \frac{(3201)(17785) - 10(5710292)}{(3201)^2 - 10(1430389)} \approx 0.0427$.

$b = \frac{1}{10}(17785 - 0.0427(3201)) \approx 1764.8.\ \therefore\ y = 0.0427\,K + 1764.8$

c) $K = 364 \Rightarrow y = 0.0427(364) + 1764.8 \approx 1780$

53. b) The table for 53, part b.

k	K_k	y_k	K_k^2	$K_k y_k$
1	1	1761	1	1761
2	75	1771	5625	132825
3	155	1772	24025	274660
4	219	1775	47961	388725
5	271	1777	73441	481567
6	351	1780	123201	624780
7	425	1783	180625	757775
8	503	1786	253009	898358
9	575	1789	330625	1028675
10	626	1791	391876	1121166
Σ	3201	17785	1430389	5710292

13.8 LAGRANGE MULTIPLIERS

1. $\nabla f = y\,\mathbf{i} + x\,\mathbf{j}$, $\nabla g = 2x\,\mathbf{i} + 4y\,\mathbf{j}$. $\nabla f = \lambda\nabla g \Rightarrow y\,\mathbf{i} + x\,\mathbf{j} = \lambda(2x\,\mathbf{i} + 4y\,\mathbf{j}) \Rightarrow y = 2x\lambda$ and $x = 4y\lambda \Rightarrow$ $\lambda = \pm\frac{\sqrt{2}}{4}$ or $x = 0$. CASE 1: If $x = 0$, then $y = 0$ but $(0,0)$ not on the ellipse. $\therefore\ x \neq 0$.
CASE 2: $x \neq 0 \Rightarrow \lambda = \pm\frac{\sqrt{2}}{4} \Rightarrow x = \pm\sqrt{2}\,y \Rightarrow (\pm\sqrt{2}\,y)^2 + 2y^2 = 1 \Rightarrow y = \pm\frac{1}{2}$. $\therefore$ f takes on its extreme values at $\left(\pm\sqrt{2}, \frac{1}{2}\right)$ and $\left(\pm\sqrt{2}, -\frac{1}{2}\right) \Rightarrow$ the extreme values of f are $\pm\frac{\sqrt{2}}{2}$.

5. $\nabla f = 2xy\,\mathbf{i} + x^2\,\mathbf{j}$, $\nabla g = \mathbf{i} + \mathbf{j}$. $\nabla f = \lambda\nabla g \Rightarrow 2xy\,\mathbf{i} + x^2\,\mathbf{j} = \lambda(\mathbf{i} + \mathbf{j}) \Rightarrow 2xy = \lambda$ and $x^2 = \lambda \Rightarrow 2xy = x^2 \Rightarrow$ $x = 0, y = 3$ or $x = 2, y = 1$. $\therefore$ f takes on its extreme values at $(0,3)$ and $(2,1)$. $\therefore$ the extreme values of f are $f(0,3) = 0$ and $f(2,1) = 4$.

9. $V = \pi r^2 h \Rightarrow 16\pi = \pi r^2 h \Rightarrow 16 = r^2 h \Rightarrow g(r,h) = r^2 h - 16$. $S = 2\pi rh + 2\pi r^2 \Rightarrow \nabla S = (2\pi h + 4\pi r)\,\mathbf{i} + 2\pi r\,\mathbf{j}$, $\nabla g = 2rh\,\mathbf{i} + r^2\,\mathbf{j}$. $\nabla S = \lambda\nabla g \Rightarrow (2\pi rh + 4\pi r)\,\mathbf{i} + 2\pi r\,\mathbf{j} = \lambda(2rh\,\mathbf{i} + r^2\,\mathbf{j}) \Rightarrow 2\pi h + 4\pi r = 2rh\lambda$ and $2\pi r = \lambda r^2 \Rightarrow$ $0 = \lambda r^2 - 2\pi r \Rightarrow r = 0$ or $\lambda = \frac{2\pi}{r}$. Now $r \neq 0 \Rightarrow \lambda = \frac{2\pi}{r} \Rightarrow 2\pi h + 4\pi r = 2rh\left(\frac{2\pi}{r}\right) \Rightarrow 2r = h \Rightarrow 16 = r^2(2r) \Rightarrow$ $r = 2 \Rightarrow h = 4$. $\therefore$ $r = 2$ cm, $h = 4$ cm give the only extreme surface area, $24\pi\ \text{cm}^2$. Since $r = 4$ cm, $h = 1$ cm $\Rightarrow$ $V = 16\pi\ \text{cm}^3$ and $S = 40\pi\ \text{cm}^2$, a larger surface area, $24\pi\ \text{cm}^2$ must be the minumum surface area.

13. $\nabla f = 2x\,\mathbf{i} + 2y\,\mathbf{j}$, $\nabla g = (2x - 2)\,\mathbf{i} + (2y - 4)\,\mathbf{j}$. $\nabla f = \lambda\nabla g \Rightarrow 2x\,\mathbf{i} + 2y\,\mathbf{j} = \lambda((2x - 2)\,\mathbf{i} + (2y - 4)\,\mathbf{j}) \Rightarrow$ $2x = \lambda(2x - 2)$ and $2y = \lambda(2y - 4) \Rightarrow x = \frac{\lambda}{\lambda - 1}$ and $y = \frac{2\lambda}{\lambda - 1}$, $\lambda \neq 1 \Rightarrow y = 2x \Rightarrow x^2 - 2x + (2x)^2 - 4(2x)$ $= 0 \Rightarrow x = 0, y = 0$ or $x = 2, y = 4$. $\therefore$ $f(0,0) = 0$ is the minimum value, $f(2,4) = 20$ is the maximum value.

17. Let $f(x,y,z) = x^2 + y^2 + z^2$ be the square of the distance to the origin. Then $\nabla f = 2x\,\mathbf{i} + 2y\,\mathbf{j} + 2z\,\mathbf{k}$, $\nabla g = y\,\mathbf{i} + x\,\mathbf{j} - \mathbf{k}$. $\nabla f = \lambda \nabla g \Rightarrow 2x\,\mathbf{i} + 2y\,\mathbf{j} + 2z\,\mathbf{k} = \lambda(y\,\mathbf{i} + x\,\mathbf{j} - \mathbf{k}) \Rightarrow 2x = \lambda y,\ 2y = \lambda x$, and $2z = -\lambda$ $\Rightarrow x = \frac{\lambda y}{2} \Rightarrow 2y = \lambda\left(\frac{\lambda y}{2}\right) \Rightarrow y = 0$ or $\lambda = \pm 2$. $y = 0 \Rightarrow x = 0 \Rightarrow -z + 1 = 0 \Rightarrow z = 1$. $\lambda = 2 \Rightarrow x = y$, $z = -1 \Rightarrow x^2 - (-1) + 1 = 0 \Rightarrow x^2 + 2 = 0$, no solution. $\lambda = -2 \Rightarrow x = -y,\ z = 1 \Rightarrow (-y)y - 1 + 1 = 0 \Rightarrow$ $y = 0$, again. $\therefore$ (0,0,1) is the point on the surface closest to the origin since this point gives the only extreme value and there is no maximum distance.

21. $f(x,y,z) = xyz$ and $g(x,y,z) = x + y + z^2 - 16 = 0 \Rightarrow \nabla f = yz\,\mathbf{i} + xz\,\mathbf{j} + xy\,\mathbf{k}$, $\nabla g = \mathbf{i} + \mathbf{j} + 2z\,\mathbf{k}$. $\nabla f = \lambda \nabla g \Rightarrow yz\,\mathbf{i} + xz\,\mathbf{j} + xy\,\mathbf{k} = \lambda(\mathbf{i} + \mathbf{j} + 2z\,\mathbf{k}) \Rightarrow yz = \lambda,\ xz = \lambda$, and $xy = 2z\lambda \Rightarrow yz = xz \Rightarrow z = 0$ or $y = x$. But $z > 0 \Rightarrow y = x \Rightarrow x^2 = 2z\lambda$ and $xz = \lambda$. Then $x^2 = 2z(xz) \Rightarrow x = 0$ or $x = 2z^2$. But $x > 0 \Rightarrow x = 2z^2 \Rightarrow y = 2z^2 \Rightarrow 2z^2 + 2z^2 + z^2 = 16 \Rightarrow z = \pm\frac{4}{\sqrt{5}}$.
Use $z = \frac{4}{\sqrt{5}}$ since $z > 0 \Rightarrow x = \frac{32}{5},\ y = \frac{32}{5}$. $f\left(\frac{32}{5}, \frac{32}{5}, \frac{4}{\sqrt{5}}\right) = \frac{4096}{25\sqrt{5}}$

25. $\nabla f = \mathbf{i} + \mathbf{j}$, $\nabla g = y\,\mathbf{i} + x\,\mathbf{j}$. $\nabla f = \lambda \nabla g \Rightarrow \mathbf{i} + \mathbf{j} = \lambda(y\,\mathbf{i} + x\,\mathbf{j}) \Rightarrow 1 = y\lambda$ and $1 = x\lambda \Rightarrow y = x \Rightarrow y^2 = 16 \Rightarrow$ $y = \pm 4 \Rightarrow x = \pm 4$. But as $x \to \infty$, $y \to 0$ and $f(x,y) \to \infty$; as $x \to -\infty$, $y \to 0$ and $f(x,y) \to -\infty$.

29. Let $g_1(x,y,z) = z - 1 = 0$ and $g_2(x,y,z) = x^2 + y^2 + z^2 - 10 = 0 \Rightarrow \nabla g_1 = \mathbf{k}$, $\nabla g_2 = 2x\,\mathbf{i} + 2y\,\mathbf{j} + 2z\,\mathbf{k}$. $\nabla f = 2xyz\,\mathbf{i} + x^2z\,\mathbf{j} + x^2y\,\mathbf{k} \Rightarrow 2xyz\,\mathbf{i} + x^2z\,\mathbf{j} + x^2y\,\mathbf{k} = \lambda(\mathbf{k}) + \mu(2x\,\mathbf{i} + 2y\,\mathbf{j} + 2z\,\mathbf{k}) \Rightarrow 2xyz = 2x\mu,\ x^2z = 2y\mu,\ x^2y = 2z + \lambda \Rightarrow xyz = x\mu \Rightarrow x = 0$ or $yz = \mu \Rightarrow \mu = y$ since $z = 1$. $x = 0$ and $z = 1 \Rightarrow y^2 - 9 = 0$ (from g_2) $\Rightarrow y = \pm 3 \Rightarrow (0, \pm 3, 1)$. $\mu = y \Rightarrow x^2z = 2y^2 \Rightarrow x^2 = 2y^2$ since $z = 1$ $\Rightarrow 2y^2 + y^2 + 1 - 10 = 0 \Rightarrow 3y^2 - 9 = 0 \Rightarrow y = \pm\sqrt{3} \Rightarrow x^2 = 2(\pm\sqrt{3})^2 = 6 \Rightarrow x =$ $\pm\sqrt{6} \Rightarrow (\pm\sqrt{6}, \pm\sqrt{3}, 1)$. $f(0,\pm 3,1) = 1$. $f(\pm\sqrt{6},\pm\sqrt{3},1) = 6(\pm\sqrt{3}) + 1 = 1 \pm 6\sqrt{3} \Rightarrow$ Maximum of f is $1 + 6\sqrt{3}$ at $(\pm\sqrt{6},\sqrt{3},1)$; minimum of f is $1 - 6\sqrt{3}$ at $(\pm\sqrt{6},-\sqrt{3},1)$

13.P PRACTICE EXERCISES

1.

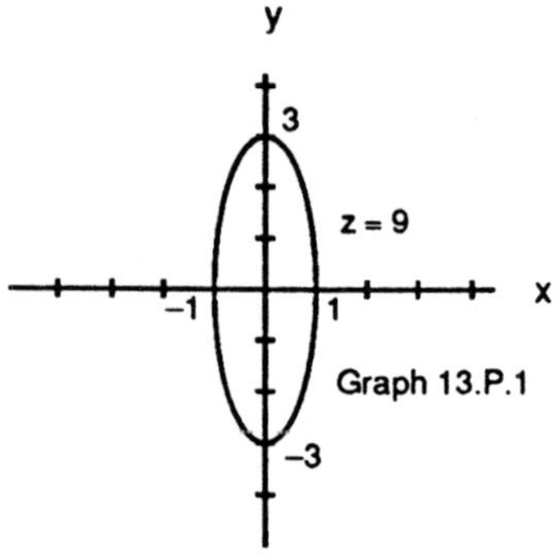

Graph 13.P.1

Domain: All points in the xy–plane

Range: $f(x,y) \geq 0$

Level curves are ellipses with major axis along the y–axis and minor axis along the x–axis.

5.

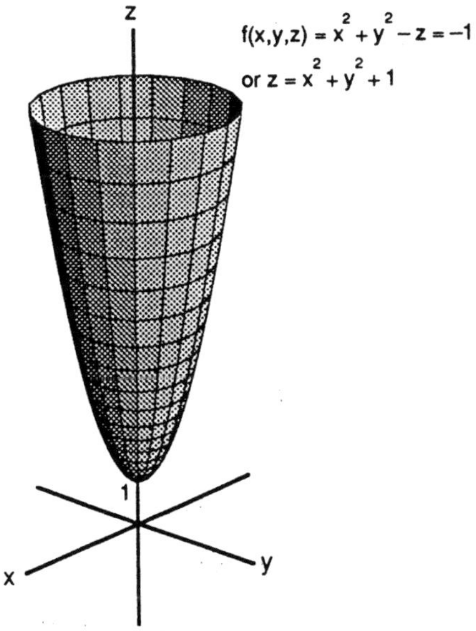

Graph 13.P.5

Domain: All (x,y,z)

Range: All Real Numbers

Level surfaces are paraboloids of revolution with the z–axis as axis.

9. $\lim_{(x,y)\to(\pi,\ln 2)} e^y \cos x = e^{\ln 2}\cos\pi = -2$

13. $\lim_{P\to(1,-1,e)} \ln|x+y+z| = \ln|1+(-1)+e| = 1$

17. $\frac{\partial g}{\partial r} = \cos\theta + \sin\theta, \frac{\partial g}{\partial\theta} = -r\sin\theta + r\cos\theta$

21. $\frac{\partial P}{\partial n} = \frac{RT}{V}, \frac{\partial P}{\partial R} = \frac{nT}{V}, \frac{\partial P}{\partial T} = \frac{nR}{V}, \frac{\partial P}{\partial V} = -\frac{nRT}{V^2}$

25. $\frac{\partial f}{\partial x} = 1 + y - 15x^2 + \frac{2x}{x^2+1}, \frac{\partial f}{\partial y} = x \Rightarrow \frac{\partial^2 f}{\partial x^2} = -30x + \frac{2-2x^2}{(x^2+1)^2}, \frac{\partial^2 f}{\partial y^2} = 0, \frac{\partial^2 f}{\partial y\partial x} = \frac{\partial^2 f}{\partial x\partial y} = 1$

29. a) $f(1,0,0) = 0, f_x(1,0,0) = y - 3z\big|_{(1,0,0)} = 0, f_y(1,0,0) = x + 2z\big|_{(1,0,0)} = 1, f_z(1,0,0) = 2y - 3x\big|_{(1,0,0)} =$
$-3 \Rightarrow L(x,y,z) = 0(x-1) + (y-0) - 3(z-0) = y - 3z.$

b) $f(1,1,0) = 1, f_x(1,1,0) = 1, f_y(1,1,0) = 1, f_z(1,1,0) = -1 \Rightarrow L(x,y,z) = 1 + (x-1) + (y-1) - 1(z-0) =$
$x + y - z - 1$

33. $dI = \frac{1}{R}dV - \frac{V}{R^2}dR \Rightarrow dI\big|_{(24,100)} = \frac{1}{100}dV - \frac{24}{100^2}dR \Rightarrow dI\big|_{dV=-1,dR=-20} = 0.038$ or increases by 0.038 amps.
% change in V $= (100)\left(-\frac{1}{24}\right) \approx -4.17\%$; % change in R $= -\frac{20}{100}(100) = -20\%$. $I = \frac{24}{100} = 0.24 \Rightarrow$ Estimated % change in I =
$\frac{dI}{I} \times 100 = \frac{0.038}{0.24} \times 100 \approx 15.83\%$

37. $\frac{\partial w}{\partial x} = 2\cos(2x-y), \frac{\partial w}{\partial y} = -\cos(2x-y), \frac{\partial x}{\partial r} = 1, \frac{\partial x}{\partial s} = \cos s, \frac{\partial y}{\partial r} = s, \frac{\partial y}{\partial s} = r \Rightarrow \frac{\partial w}{\partial r} = 2\cos(2x-y)(1) +$
$(-\cos(2x-y)(s)) = 2\cos(2r + 2\sin s - rs) - s\cos(2r + 2\sin s - rs) \Rightarrow \frac{\partial w}{\partial r}\big|_{r=\pi,s=0} = 2.$
$\frac{\partial w}{\partial s} = 2\cos(2x-y)(\cos s) + (-\cos(2x-y)(r)) = 2\cos(2r + 2\sin s - rs)(\cos s) - r\cos(2r + 2\sin s - rs) \Rightarrow$
$\frac{\partial w}{\partial s}\big|_{r=\pi,s=0} = 2 - \pi$

41. $\frac{\partial f}{\partial x} = y + z, \frac{\partial f}{\partial y} = x + z, \frac{\partial f}{\partial z} = y + x, \frac{dx}{dt} = -\sin t, \frac{dy}{dt} = \cos t, \frac{dz}{dt} = -2\sin 2t \Rightarrow \frac{df}{dt} = -(y+z)\sin t + (x+z)\cos t$
$- 2(y+x)\sin 2t = -(\sin t + \cos 2t)\sin t + (\cos t + \cos 2t)\cos t - 2(\sin t + \cos t)\sin 2t \Rightarrow \frac{df}{dt}\big|_{t=1} =$
$-(\sin 1 + \cos 2)\sin 1 + (\cos 1 + \cos 2)\cos 1 - 2(\sin 1 + \cos 1)\sin 2$

45. $\nabla f = \left(\frac{2}{2x+3y+6z}\right)\mathbf{i} + \left(\frac{3}{2x+3y+6z}\right)\mathbf{j} + \left(\frac{6}{2x+3y+6z}\right)\mathbf{k} \Rightarrow \nabla f\big|_{(-1,-1,1)} = 2\,\mathbf{i} + 3\,\mathbf{j} + 6\,\mathbf{k}.$
$\mathbf{u} = \frac{\nabla f}{|\nabla f|} = \frac{2\,\mathbf{i} + 3\,\mathbf{j} + 6\,\mathbf{k}}{\sqrt{2^2 + 3^2 + 6^2}} = \frac{2}{7}\mathbf{i} + \frac{3}{7}\mathbf{j} + \frac{6}{7}\mathbf{k}$. f increases most rapidly in the direction $\mathbf{u} = \frac{2}{7}\mathbf{i} + \frac{3}{7}\mathbf{j} + \frac{6}{7}\mathbf{k}$; decreases most rapidly in the direction $-\mathbf{u} = -\frac{2}{7}\mathbf{i} - \frac{3}{7}\mathbf{j} - \frac{6}{7}\mathbf{k}$. $\left(D_{\mathbf{u}}f\right)_{P_0} = \nabla f \cdot \mathbf{u} = 7,$
$\left(D_{-\mathbf{u}}f\right)_{P_0} = -7$. $\mathbf{u}_1 = \frac{\mathbf{A}}{|\mathbf{A}|} = \frac{2}{7}\mathbf{i} + \frac{3}{7}\mathbf{j} + \frac{6}{7}\mathbf{k}$ since $\mathbf{A} = \nabla f. \Rightarrow \left(D_{\mathbf{u}_1}f\right)_{P_0} = 7.$

49. $\nabla f = 2x\,\mathbf{i} - \mathbf{j} - 5\,\mathbf{k} \Rightarrow \nabla f\big|_{(2,-1,1)} = 4\,\mathbf{i} - \mathbf{j} - 5\,\mathbf{k} \Rightarrow$ Tangent Plane: $4(x-2) - (y+1) - 5(z-1) = 0 \Rightarrow$
$4x - y - 5z = 4$; Normal Line: $x = 2 + 4t,\ y = -1 - t,\ z = 1 - 5t$

53.

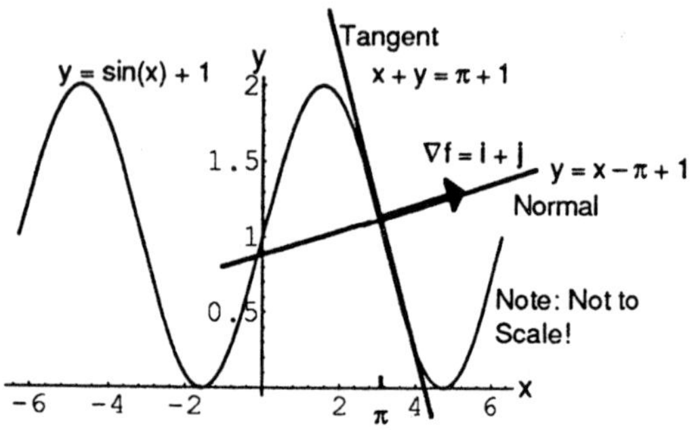

Graph 13.P.53

$\nabla f = (-\cos x)\,\mathbf{i} + \mathbf{j} \Rightarrow \nabla f\big|_{(\pi,1)} = \mathbf{i} + \mathbf{j} \Rightarrow$ Tangent Line: $(x - \pi) + (y - 1) = 0 \Rightarrow x + y = \pi + 1$;
Normal Line: $y - 1 = 1(x - \pi) \Rightarrow y = x - \pi + 1$

57. $f(x,y,z) = xyz \Rightarrow \nabla f = yz\,\mathbf{i} + xz\,\mathbf{j} + xy\,\mathbf{k}$. At $(1,1,1)$, $\nabla f = \mathbf{i} + \mathbf{j} + \mathbf{k} \Rightarrow$ the maximum value of $D_{\mathbf{u}}f\big|_{(1,1,1)} = |\nabla f| = \sqrt{3}$.

61. $f_x(x,y) = 6x^2 + 3y = 0$ and $f_y(x,y) = 3x + 6y^2 = 0 \Rightarrow x = 0, y = 0$ or $x = -\frac{1}{2}, y = -\frac{1}{2} \Rightarrow$ critical points are $(0,0)$ and $\left(-\frac{1}{2}, -\frac{1}{2}\right)$ For $(0,0)$: $f_{xx}(0,0) = 12x\big|_{(0,0)} = 0$, $f_{yy}(,0) = 6y\big|_{(0,0)} = 0$, $f_{xy}(0,0) = 3 \Rightarrow$
$f_{xx}f_{yy} - f_{xy}^{\,2} = -9 \Rightarrow$ Saddle Point. $f(0,0) = 0$. For $\left(-\frac{1}{2}, -\frac{1}{2}\right)$, $f_{xx} = -6$, $f_{yy} = -6$, $f_{xy} = 3 \Rightarrow f_{xx}f_{yy} - f_{xy}^{\,2} = 27 > 0$
and $f_{xx} < 0 \Rightarrow$ Maximum. $f\left(-\frac{1}{2}, -\frac{1}{2}\right) = \frac{1}{4}$

65.

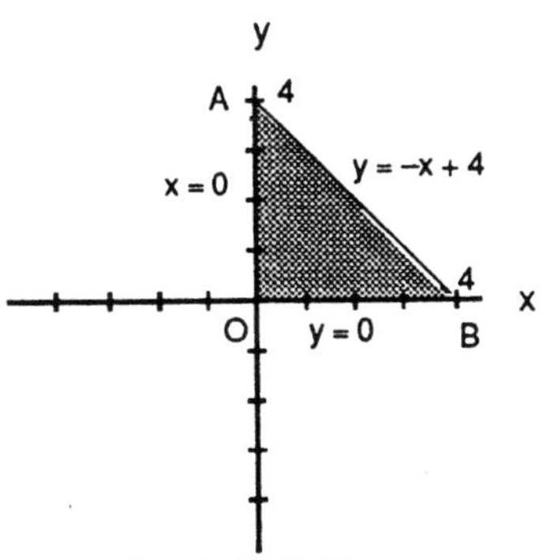

Graph 13.P.65

1. On OA, $f(x,y) = y^2 + 3y = f(0,y)$ for $0 \le y \le 4$. $f(0,0) = 0$, $f(0,4) = 28$. $f'(0,y) = 2y + 3 = 0 \Rightarrow y = -\frac{3}{2}$. But $\left(0, -\frac{3}{2}\right)$ is not in the region.
2. On AB, $f(x,y) = x^2 - 10x + 28 = f(x, -x + 4)$ for $0 \le x \le 4$. $f(4,0) = 4$. $f'(x, -x + 4) = 2x - 10 = 0 \Rightarrow x = 5$, $y = -1$. But $(5,-1)$ not in the region.
3. On OB, $f(x,y) = x^2 - 3x = f(x,0)$ for $0 \le x \le 4$. $f'(x,0) = 2x - 3 \Rightarrow x = \frac{3}{2}$, $y = 0 \Rightarrow \left(\frac{3}{2}, 0\right)$ is a critical point. $f\left(\frac{3}{2}, 0\right) = -\frac{9}{4}$

65. (Continued)

4. For the interior of the triangular region, $f_x(x,y) = 2x + y - 3 = 0$ and $f_y(x,y) = x + 2y + 3 = 0 \Rightarrow x = 3$, $y = -3$. But $(3,-3)$ is not in the region.

$\therefore$ the absolute maximum is 28 at (0,4); the absolute minimum is $-\frac{9}{4}$ at $\left(\frac{3}{2}, 0\right)$.

69.

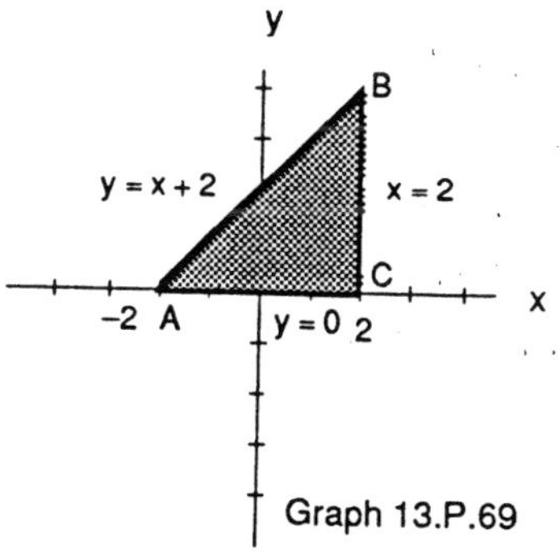

Graph 13.P.69

1. On AB, $f(x,y) = -2x + 4 = f(x, x + 2)$ for $-2 \le x \le 2$. $f(-2,0) = 8$, $f(2,4) = 0$. $f'(x, x + 2) = -2 \Rightarrow$ no critical points in the interior of AB.
2. On BC, $f(x,y) = -y^2 + 4y = f(2,y)$ for $0 \le y \le 4$. $f(2,0) = 0$. $f'(2,y) = -2y + 4 = 0 \Rightarrow y = 2$, $x = 2 \Rightarrow (2,2)$ is a critical point. $f(2,2) = 4$.
3. On AC, $f(x,y) = x^2 - 2x = f(x,0)$ for $-2 \le x \le 2$. $f'(x,0) = 2x - 2 = 0 \Rightarrow x = 1$, $y = 0 \Rightarrow (1,0)$ is a critical point. $f(1,0) = -1$.
4. For the interior of the triangular region, $f_x(x,y) = 2x - 2 = 0$ and $f_y(x,y) = -2y + 4 = 0 \Rightarrow x = 1$, $y = 2 \Rightarrow (1,2)$ is a critical point $f(1,2) = 3$.

$\therefore$ the absolute maximum is 8 at $(-2,0)$; the absolute minimum is -1 at $(1,0)$.

73. $\nabla T = 400yz^2\,\mathbf{i} + 400xz^2\,\mathbf{j} + 800xyz\,\mathbf{k}$, $\nabla g = 2x\,\mathbf{i} + 2y\,\mathbf{j} + 2z\,\mathbf{k}$. $\nabla T = \lambda \nabla g \Rightarrow 400yz^2\,\mathbf{i} + 400xz^2\,\mathbf{j} + 800xyz\,\mathbf{k} = \lambda(2x\,\mathbf{i} + 2y\,\mathbf{j} + 2z\,\mathbf{k}) \Rightarrow 400yz^2 = 2x\lambda$, $400xz^2 = 2y\lambda$, and $800xyz = 2z\lambda$. Solving this system yields the following points: $(0,\pm1,0)$, $(\pm1,0,0)$, $\left(\pm\frac{1}{2}, \pm\frac{1}{2}, \pm\frac{\sqrt{2}}{2}\right)$. $T(0,\pm1,0) = 0$, $T(\pm1,0,0) = 0$, $T\left(\pm\frac{1}{2}, \pm\frac{1}{2}, \pm\frac{\sqrt{2}}{2}\right) = \pm50$. $\therefore$ 50 is the maximum at $\left(\frac{1}{2}, \frac{1}{2}, \pm\frac{\sqrt{2}}{2}\right)$ and $\left(-\frac{1}{2}, -\frac{1}{2}, \pm\frac{\sqrt{2}}{2}\right)$; -50 is the minimum at $\left(\frac{1}{2}, -\frac{1}{2}, \pm\frac{\sqrt{2}}{2}\right)$ and $\left(-\frac{1}{2}, \frac{1}{2}, \pm\frac{\sqrt{2}}{2}\right)$.

CHAPTER 14

MULTIPLE INTEGRALS

14.1 DOUBLE INTEGRALS

1. $\int_0^3 \int_0^2 \left(4 - y^2\right) dy\, dx = \int_0^3 \left[4y - \frac{y^3}{3}\right]_0^2 dx = \frac{16}{3}\int_0^3 dx = 16$

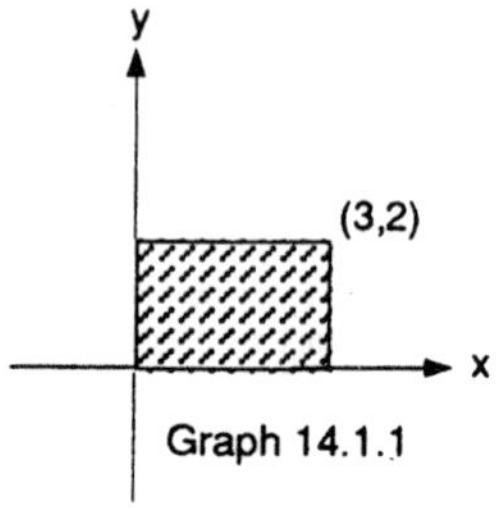

Graph 14.1.1

5. $\int_0^{\pi} \int_0^{x} (x \sin y)\, dy\, dx = \int_0^{\pi} [-x \cos y]_0^x\, dx = \int_0^{\pi} (x - x \cos x)\, dx = \frac{\pi^2}{2} + 2$

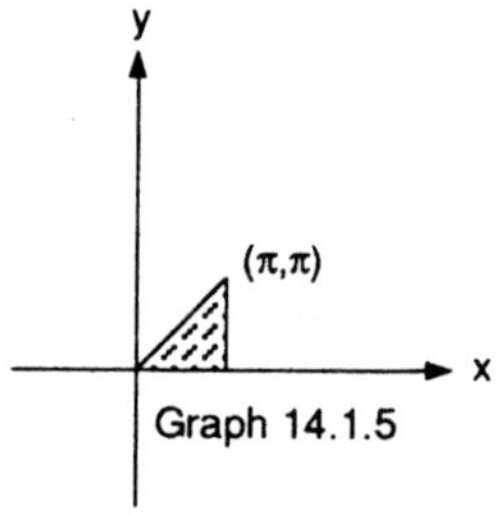

Graph 14.1.5

9. $\int_0^1 \int_0^{y^2} 3y^3 e^{xy}\, dx\, dy = \int_0^1 \left[3y^2 e^{xy}\right]_0^{y^2} dy =$

$\int_0^1 \left(3y^2 e^{y^3} - 3y^2\right) dy = \left[e^{y^3} - y^3\right]_0^1 = e - 2$

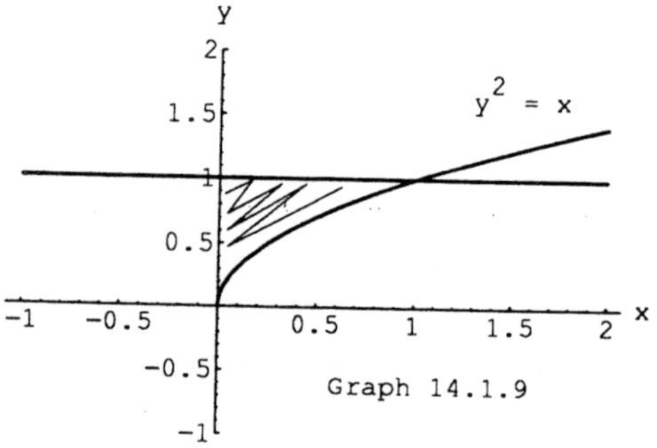

Graph 14.1.9

13. $\int_0^1 \int_0^{1-y} \left(x^2 + y^2\right) dx\, dy = \int_0^1 \left[\frac{x^3}{3} + y^2 x\right]_0^{1-y} dy = \int_0^1 \left[\frac{(1-y)^3}{3} + y^2 - y^3\right] dy = \frac{1}{6}$

17. $\int_{-2}^{0} \int_{v}^{-v} 2\, dp\, dv = \int_{-2}^{0} [p]_v^{-v}\, dv =$

$2\int_{-2}^{0} -2\, v\, dv = -2\left[2\, v^2\right]_{-2}^{0} = 8$

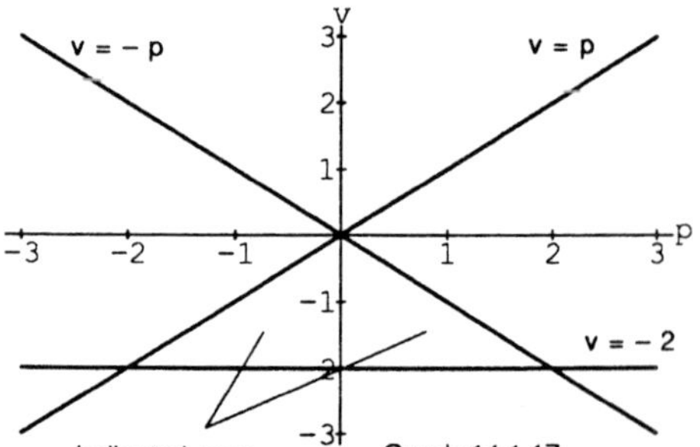

Graph 14.1.17

21. $\displaystyle\int_2^4 \int_0^{(4-y)/2} dx\ dy$

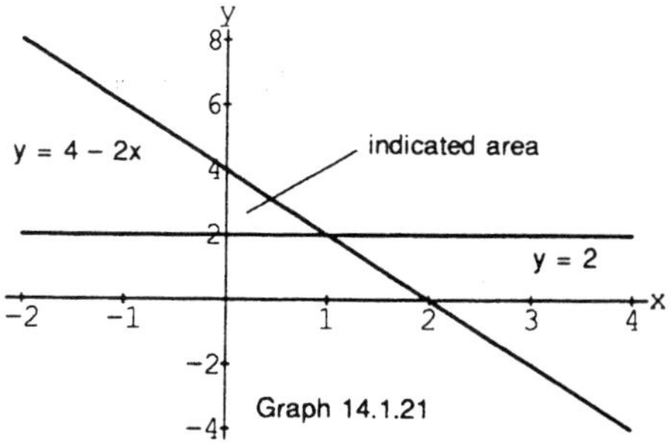

Graph 14.1.21

25. $\displaystyle\int_1^e \int_{\ln y}^1 dx\ dy$

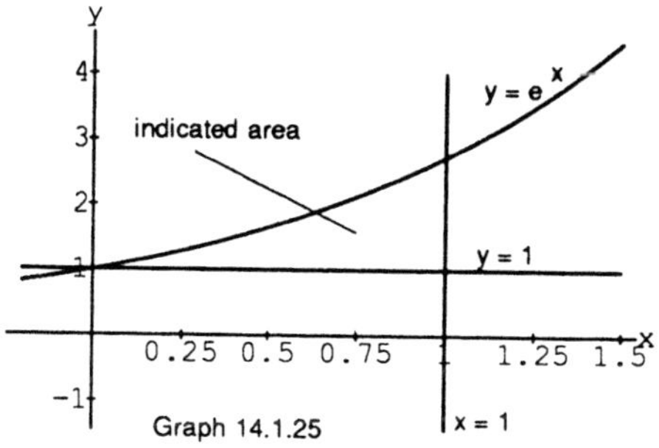

Graph 14.1.25

29. $\displaystyle\int_{-1}^1 \int_0^{\sqrt{1-x^2}} 3y\ dy\ dx$

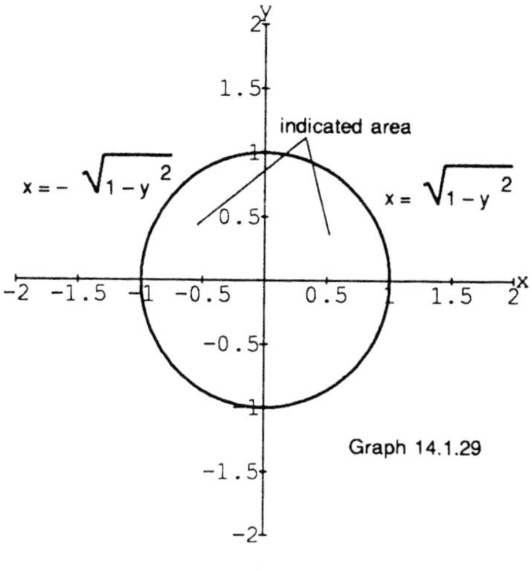

Graph 14.1.29

33. $\displaystyle\int_0^2 \int_x^2 2y^2 \sin xy\, dy\, dx = \int_0^2 \int_0^y 2y^2 \sin xy\, dx\, dy =$
$\displaystyle\int_0^2 [-2y\cos xy]_0^y\, dy = \int_0^2 \left(-\cos y^2(2y) + 2y\right) dy = 4 - \sin 4$

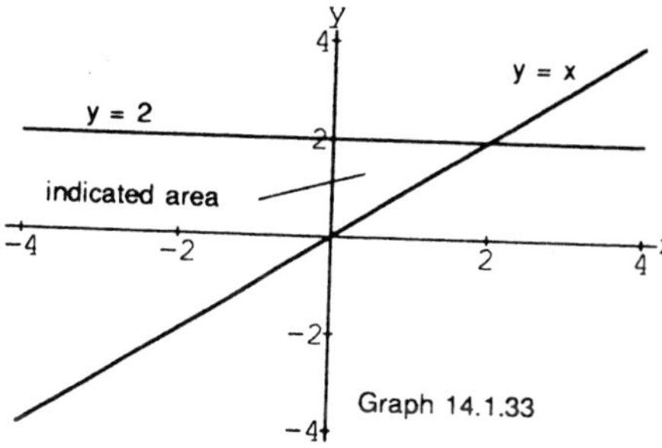

Graph 14.1.33

37. $\displaystyle\int_0^3 \int_{\sqrt{x/3}}^1 e^{y^3}\,dy\,dx = \int_0^1 \int_0^{3y^2} e^{y^3}\,dx\,dy =$
$\displaystyle\int_0^1 e^{y^3}\left(3y^2\right) dy = \left[e^{y^3}\right]_0^1 = e - 1$

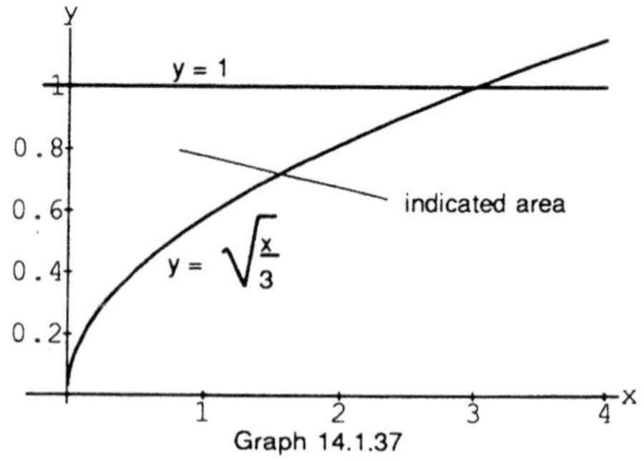

Graph 14.1.37

41. $\displaystyle V = \int_0^2 \int_0^3 \left(4 - y^2\right) dx\,dy = \int_0^2 \left[4x - y^2x\right]_0^3 dy = \int_0^2 \left(12 - 3y^2\right) dy = 16$

45. To maximize the integral, we want the domain to include all points where the integrand is positive and to exclude all points where the integrand is negative. These criteria are met by the set of points (x, y) such that $4 - x^2 - 2y^2 \geq 0$ or $x^2 + 2y^2 \leq 4$. It consists of the ellipse $x^2 + 2y^2 = 4$ and its interior.

49. $\displaystyle\int_1^3 \int_1^x \frac{1}{xy}\,dy\,dx = 0.603$

14.2 AREAS, MOMENTS, AND CENTERS OF MASS

1. $\displaystyle\int_0^2 \int_0^{2-x} dy\,dx = \int_0^2 (2 - x)\,dx = 2$ or
$\displaystyle\int_0^2 \int_0^{2-y} dx\,dy = \int_0^2 (2 - y)\,dx = 2$

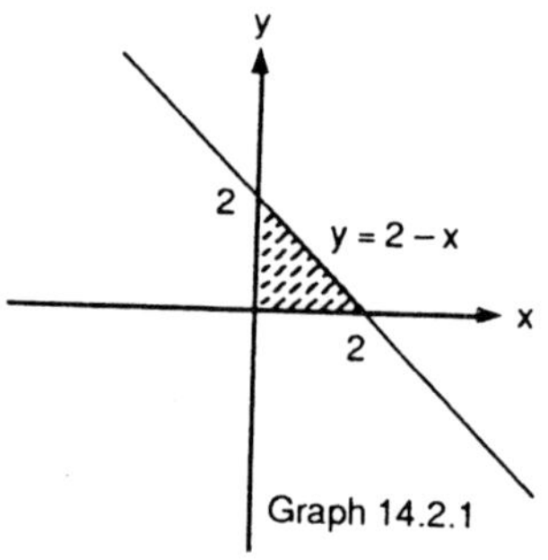

Graph 14.2.1

5. $\displaystyle\int_0^{\ln 2} \int_0^{e^x} dy\,dx = \int_0^{\ln 2} e^x\,dx = 1$

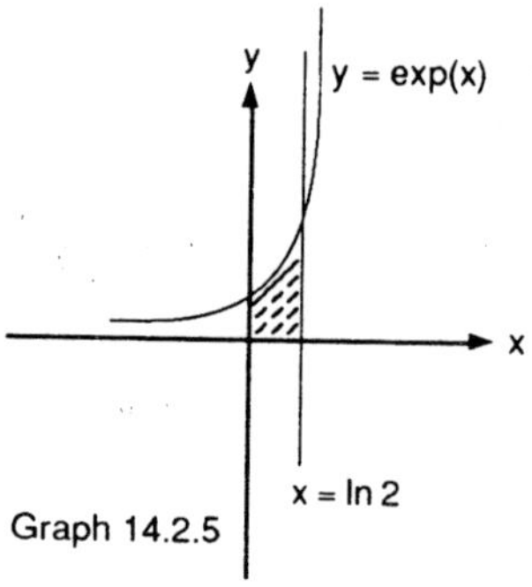

Graph 14.2.5

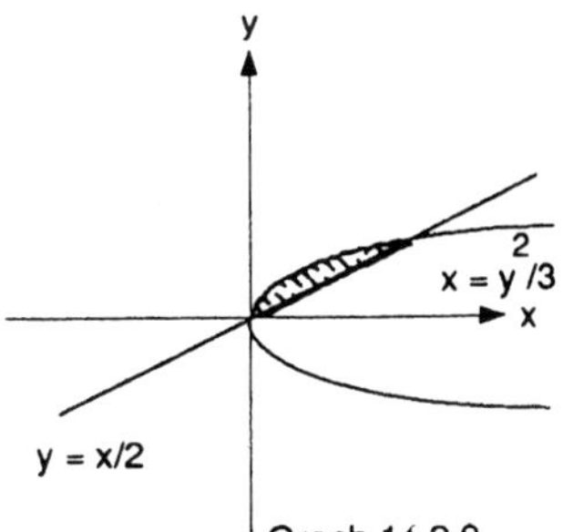

Graph 14.2.9

9. $$\int_0^6 \int_{y^2/3}^{2y} dx\,dy = \int_0^6 \left(2y - y^2/3\right) dy = 12$$

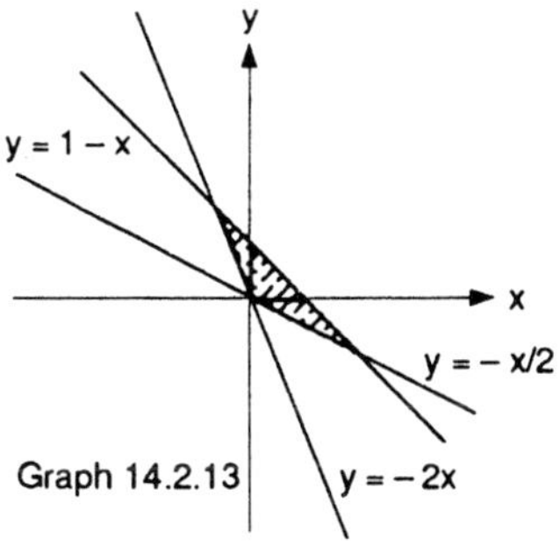

Graph 14.2.13

13. $$\int_{-1}^{0} \int_{-2x}^{1-x} dy\,dx + \int_0^2 \int_{-x/2}^{1-x} dy\,dx = \int_{-1}^{0} (1 + x)\,dx + \int_0^2 (1 - x/2)\,dx = \frac{3}{2}$$

17. Average height $= \frac{1}{4}\int_0^2 \int_0^2 \left(x^2 + y^2\right) dy\,dx = \frac{1}{4}\int_0^2 \left[x^2 y + \frac{y^3}{3}\right]_0^2 dx = \frac{1}{4}\int_0^2 \left(2x^2 + \frac{8}{3}\right) dx = \frac{1}{2}\left[\frac{x^3}{3} + \frac{4x}{3}\right]_0^2 = \frac{8}{3}$

21. $$M = \int_0^2 \int_{y^2/2}^{4-y} dx\,dy = \int_0^2 \left(4 - y - \frac{y^2}{2}\right) dy = \frac{14}{3}$$

$$M_y = \int_0^2 \int_{y^2/2}^{4-y} x\,dx\,dy = \frac{1}{2}\int_0^2 \left[x^2\right]_{y^2/2}^{4-y} dx = \frac{1}{2}\int_0^2 \left(16 - 8y + y^2 - \frac{y^4}{4}\right) dy = \frac{128}{15}$$

$$M_x = \int_0^2 \int_{y^2/2}^{4-y} y\,dx\,dy = \int_0^2 y\left(4 - y - \frac{y^2}{2}\right) dy = \frac{10}{3} \Rightarrow \overline{x} = \frac{64}{35},\ \overline{y} = \frac{5}{7}$$

25. $$M = \int_0^a \int_0^{\sqrt{a^2-x^2}} dy\,dx = \frac{\pi a^2}{4};\ M_y = \int_0^a \int_0^{\sqrt{a^2-x^2}} x\,dy\,dx =$$

$$\int_0^a [xy]_0^{\sqrt{a^2-x^2}} dx = -\frac{1}{2}\int_0^a \sqrt{a^2 - y^2}(-2x)\,dx = \frac{a^3}{3} \Rightarrow \overline{x} = \overline{y} = \frac{4a}{3\pi},\ \text{by symmetry}$$

29. $$M = \int_{-\infty}^{0} \int_0^{\exp(x)} dy\,dx = \int_{-\infty}^{0} e^x\,dx = \lim_{t \to -\infty} \int_t^0 e^x\,dx = 1$$

$$M_y = \int_{-\infty}^{0} \int_0^{\exp(x)} x\,dy\,dx = \int_{-\infty}^{0} x\,e^x\,dx = \lim_{t \to -\infty} \int_t^0 x\,e^x\,dx = -1$$

$$M_x = \int_{-\infty}^{0} \int_0^{\exp(x)} y\,dy\,dx = \frac{1}{2}\int_{-\infty}^{0} e^{2x}\,dx = \frac{1}{2}\lim_{t \to -\infty} \int_t^0 e^{2x}\,dx = \frac{1}{4} \Rightarrow \overline{x} = -1,\ \overline{y} = \frac{1}{4}$$

33. $M = \int_0^1 \int_x^{2-x} (6x + 3y + 3)\,dy\,dx = \int_0^1 \left(-12x^2 + 12\right) dx = 8;\ M_y = \int_0^1 \int_x^{2-x} x(6x + 3y + 3)\,dy\,dx =$
$\int_0^1 \left(12x - 12x^3\right) dx = 3\ ;\ M_x = \int_0^1 \int_x^{2-x} y(6x + 3y + 3)\,dy\,dx = \int_0^1 \left(14 - 6x - 6x^2 - 2x^3\right) dx = \frac{17}{2} \Rightarrow$
$\overline{x} = \frac{3}{8},\ \overline{y} = \frac{17}{16}$

37. $M = \int_{-1}^1 \int_0^{x^2} (7y + 1)\,dy\,dx = \int_{-1}^1 \left(\frac{7x^4}{2} + x^2\right) dx = \frac{31}{15}$
$M_x = \int_{-1}^1 \int_0^{x^2} y(7y + 1)\,dy\,dx = \int_{-1}^1 \left(\frac{7x^6}{3} + \frac{x^4}{2}\right) dx = \frac{13}{15},\ M_y = \int_{-1}^1 \int_0^{x^2} x(7y + 1)\,dy\,dx = \int_{-1}^1 \left(\frac{7x^5}{2} + x^3\right) dx = 0$
$\therefore\ \overline{x} = 0,\ \overline{y} = \frac{13}{31},\ I_y = \int_{-1}^1 \int_0^{x^2} x^2(7y + 1)\,dy\,dx = \int_{-1}^1 x^2\left(\frac{7x^4}{2} + x^2\right) dx = \frac{7}{5} \Rightarrow R_y = \sqrt{\frac{I_y}{M}} = \sqrt{\frac{21}{31}}$

41. $M = \int_{-1}^1 \int_0^{a(1-x^2)} dy\,dx = 2a\int_0^1 \left(1 - x^2\right) dx = 2a\left[x - \frac{x^3}{3}\right]_0^1 = \frac{4a}{3},\ M_x = \int_{-1}^1 \int_0^{a(1-x^2)} y\,dy\,dx =$
$\frac{2a^2}{2}\int_0^1 \left(1 - 2x^2 + x^4\right) dx = a^2\left[x - \frac{2x^3}{3} + \frac{x^5}{5}\right]_0^1 = \frac{8a^2}{15}\ \therefore\ \overline{y} = \frac{M_x}{M} = \frac{8a^2/15}{4a/3} = \frac{2a}{5}$; The angle formed by the c. m., the fulcrum and origin plus 45° must remain less than 90°. i. e. $\tan^{-1}\left(\frac{2a}{5}\right) + \frac{\pi}{4} < \frac{\pi}{2} \Rightarrow 0 < a < \frac{5}{2}$.

14.3 DOUBLE INTEGRALS IN POLAR FORM

1. $\int_{-1}^1 \int_0^{\sqrt{1-x^2}} dy\,dx = \int_0^\pi \int_0^1 r\,dr\,d\theta = \frac{1}{2}\int_0^\pi d\theta = \frac{\pi}{2}$

5. $\int_{-a}^a \int_{-\sqrt{a^2-x^2}}^{\sqrt{a^2-x^2}} dy\,dx = \int_0^{2\pi} \int_0^a r\,dr\,d\theta = \frac{a^2}{2}\int_0^{2\pi} d\theta = \pi a^2$

9. $\int_{-1}^0 \int_{-\sqrt{1-x^2}}^0 \frac{2}{1 + \sqrt{x^2 + y^2}}\,dy\,dx = \int_\pi^{3\pi/2} \int_0^1 \frac{2r}{1 + r}\,dr\,d\theta = 2\int_\pi^{3\pi/2} (1 - \ln 2)\,d\theta = (1 - \ln 2)\pi$

13. $\int_0^2 \int_0^{\sqrt{1 - (x-1)^2}} 3xy\,dy\,dx = \int_0^{\pi/2} \int_0^{2\cos\theta} 3(r\cos\theta)(r\sin\theta)r\,dr\,d\theta = 12\int_0^{\pi/2} \cos^5\theta\sin\theta\,d\theta =$
$-2\left[\frac{\cos^6\theta}{6}\right]_0^{\pi/2} = 2$

17. $\int_0^{\pi/2} \int_0^{2\sqrt{2-\sin 2\theta}} r\,dr\,d\theta = 2\int_0^{\pi/2} \left(2 - \sin 2\theta\right) d\theta = 2(\pi - 1)$

21. $A = \int_0^{\pi/2} \int_0^{1+\sin\theta} r\,dr\,d\theta = \frac{1}{2}\int_0^{\pi/2}\left(\frac{3}{2} + 2\sin\theta - \frac{\cos 2\theta}{2}\right)d\theta = \frac{3\pi + 8}{8}$

25. $M_x = \int_0^{\pi}\int_0^{1-\cos\theta} 3r^2 \sin\theta\ dr\,d\theta = \int_0^{\pi}\left(1 - \cos\theta\right)^3 \sin\theta\,d\theta = 4$

29. $M = 2\int_0^{\pi}\int_0^{1+\cos\theta} r\,dr\,d\theta = \int_0^{\pi}\left(1 + \cos\theta\right)^2 d\theta = \frac{3\pi}{2}$, $M_y = \int_0^{2\pi}\int_0^{1+\cos\theta} \cos\theta\, r^2\,dr\,d\theta =$
$\int_0^{2\pi}\left(\frac{4\cos\theta}{3} + \frac{15}{24} + \cos 2\theta - \sin^2\theta\cos\theta + \frac{\cos 4\theta}{4}\right)d\theta = \frac{5\pi}{4} \Rightarrow \overline{x} = \frac{5}{6}$, $\overline{y} = 0$ by symmetry

33. $\text{Average} = \frac{1}{\pi a^2}\int_{-a}^{a}\int_{-\sqrt{a^2-x^2}}^{\sqrt{a^2-x^2}} \sqrt{x^2+y^2}\,dy\,dx = \frac{1}{\pi a^2}\int_0^{2\pi}\int_0^{a} r^2\,dr\,d\theta = \frac{a}{3\pi}\int_0^{2\pi} d\theta = \frac{2a}{3}$

14.4 TRIPLE INTEGRALS IN RECTANGULAR COORDINATES

1. $\int_0^1\int_0^{1-z}\int_0^2 dx\,dy\,dz = 2\int_0^1\int_0^{1-z} dy\,dz = 2\int_0^1 1 - z\,dz = 1$

5. $\int_0^1\int_0^1\int_0^1 \left(x^2+y^2+z^2\right) dz\,dy\,dx = \int_0^1\int_0^1 \left(x^2+y^2+\frac{1}{3}\right) dy\,dx = \int_0^1\left(x^2+\frac{2}{3}\right) dx = 1$

9. $\int_0^1\int_0^{\pi}\int_0^{\pi} y\sin z\,dx\,dy\,dz = \int_0^1\int_0^{\pi} \pi y\sin z\,dy\,dz = \frac{\pi^3}{2}\int_0^1 \sin z\,dz = \frac{\pi^3}{2}(1 - \cos 1)$

13. $\int_0^1\int_0^{2-x}\int_0^{2-x-y} dz\,dy\,dx = \int_0^1\int_0^{2-x} (2 - x - y)\,dy\,dx = \int_0^1\left(\frac{x^2}{2} - 2x + 2\right)dx = \frac{7}{6}$

17. $\int_0^{\pi/4}\int_0^{\ln\sec v}\int_{-\infty}^{2t} e^x\,dx\,dt\,dv = \int_0^{\pi/4}\int_0^{\ln\sec v} e^{2t}\,dt\,dv = \int_0^{\pi/4}\frac{\sec^2 v}{2}\,dv = \left[\frac{\tan v}{2}\right]_0^{\pi/4} = \frac{1}{2} - \frac{\pi}{8} = \frac{4-\pi}{8}$

21. $V = \int_0^1\int_{-1}^1\int_0^{y^2} dz\,dy\,dx = \int_0^1\int_{-1}^1 y^2\,dy\,dx = \frac{2}{3}\int_0^1 dx = \frac{2}{3}$

25. $V = \int_0^1\int_0^{2-2x}\int_0^{3-3x-3y/2} dz\,dy\,dx = \int_0^1\int_0^{2-2x}\left(3 - 3x - \frac{3}{2}y\right) dy\,dx = \int_0^1\left(3 - 6x + 3x^2\right) dx = 1$

29. $V = \int_0^4\int_0^{\left(\sqrt{16-y^2}\right)/2}\int_0^{4-y} dx\,dz\,dy = \int_0^4\int_0^{\left(\sqrt{16-y^2}\right)/2} (4 - y)\,dz\,dy = \int_0^4 \frac{\sqrt{16-y^2}}{2}(4 - y)\,dy = 8\pi - \frac{32}{3}$

33. $\text{average} = \int_0^1\int_0^1\int_0^1 \left(x^2+y^2+z^2\right) dz\,dy\,dx = \int_0^1\int_0^1\left(x^2+y^2+\frac{1}{3}\right) dy\,dx = \int_0^1\left(x^2+\frac{2}{3}\right) dx = 1$

14.5 MASSES AND MOMENTS IN THREE DIMENSIONS

1. $I_x = \int_{-c/2}^{c/2}\int_{-b/2}^{b/2}\int_{-a/2}^{a/2}\left(y^2+z^2\right)dx\,dy\,dz = 4a\int_0^{c/2}\int_0^{b/2}\left(y^2+z^2\right)dy\,dz = 4a\int_0^{c/2}\left(\frac{b^3}{24}+\frac{z^2b}{2}\right)dz =$
$\frac{abc}{12}\left(b^2+c^2\right) \Rightarrow I_x = \frac{M}{12}\left(b^2+c^2\right);\ R_x = \sqrt{\frac{b^2+c^2}{12}},\ R_y = \sqrt{\frac{a^2+c^2}{12}},\ R_z = \sqrt{\frac{a^2+b^2}{12}}$

5. $M = 4\int_0^1\int_0^1\int_{4y^2}^4 dz\,dy\,dx = 4\int_0^1\int_0^1\left(4-4y^2\right)dy\,dx = 16\int_0^1 \frac{2}{3}dx = \frac{32}{3}$

$M_{xy} = 4\int_0^1\int_0^1\int_{4y^2}^4 z\,dz\,dy\,dx = 2\int_0^1\int_0^1\left(16-16y^4\right)dy\,dx = \frac{128}{5}\int_0^1 dx = \frac{128}{5} \Rightarrow \overline{z} = \frac{12}{5},\ \overline{x} = \overline{y} = 0$ by
symmetry; $I_z = 4\int_0^1\int_0^1\int_{4y^2}^4\left(x^2+y^2\right)dz\,dy\,dx = 16\int_0^1\int_0^1\left(x^2-x^2y^2+y^2-y^4\right)dy\,dx = 16\int_0^1\left(\frac{2x^2}{3}+\frac{2}{15}\right)dx =$
$\frac{256}{45}$; $I_x = 4\int_0^1\int_0^1\int_{4y^2}^4\left(y^2+z^2\right)dz\,dy\,dx = 4\int_0^1\int_0^1\left[\left(4y^2+\frac{64}{3}\right)-\left(4y^4+\frac{64y^6}{3}\right)\right]dy\,dx = 4\int_0^1\frac{1976}{105}dx =$
$\frac{7904}{105}$; $I_y = 4\int_0^1\int_0^1\int_{4y^2}^4\left(x^2+z^2\right)dz\,dy\,dx = 4\int_0^1\int_0^1\left[\left(4x^2+\frac{64}{3}\right)-\left(4x^2y^2+\frac{64y^6}{3}\right)\right]dy\,dx =$
$4\int_0^1\left(\frac{8}{3}x^2+\frac{128}{7}\right)dx = \frac{4832}{63}$

$c^2 = 8 \Rightarrow c = 2\sqrt{2}$, since $c > 0$

9. $I_L = \int_{-2}^{2}\int_{-2}^{4}\int_{-1}^{1-y/2}\left((y-6)^2+z^2\right)dz\,dy\,dx = \int_{-2}^{2}\int_{-2}^{4}\left(\frac{(y-6)^2(4-y)}{2}+\frac{(2-y)^3}{24}+\frac{1}{3}\right)dy\,dx =$
$4\int_{-2}^{4}\left[\frac{13t^3}{24}+5t^2+16t+\frac{49}{3}\right]dt = 1386$, where $t = 2-y$; $M = 36$, $R_L = \sqrt{\frac{I_L}{M}} = \sqrt{\frac{77}{2}}$

$R_L = \sqrt{\frac{I_L}{M}} = \sqrt{\frac{5}{3}}$

13. $M = \int_0^2\int_0^{2-x}\int_0^{2-x-y} 2x\,dz\,dy\,dx = \int_0^2\int_0^{2-x}\left(4x-2x^2-2xy\right)dy\,dx = \int_0^2\left(x^3-4x^2+4x\right)dx = \frac{4}{3}$

$M_{xy} = \int_0^2\int_0^{2-x}\int_0^{2-x-y} 2xz\,dz\,dy\,dx = \int_0^2\int_0^{2-x} x(2-x-y)^2\,dy\,dx = \int_0^2 \frac{x(2-x)^3}{3}dx = \frac{8}{15}$;
$M_{xz} = \frac{8}{15}$ by symmetry; $M_{yz} = \int_0^2\int_0^{2-x}\int_0^{2-x-y} 2x^2\,dz\,dy\,dx = \int_0^2\int_0^{2-x} 2x^2(2-x-y)\,dy\,dx =$
$\int_0^2\left(2x-x^2\right)^2 dx = \frac{16}{15} \Rightarrow \overline{x} = \frac{4}{5},\ \overline{y} = \overline{z} = \frac{2}{5}$

15. $M = \int_0^1 \int_0^1 \int_0^1 (x + y + z + 1)\, dz\, dy\, dx = \int_0^1 \int_0^1 \left(x + y + \frac{3}{2}\right) dy\, dx = \int_0^1 (x + 2)\, dx = \frac{5}{2}$

$M_{xy} = \int_0^1 \int_0^1 \int_0^1 (x + y + z + 1)z\, dz\, dy\, dx = \frac{1}{2}\int_0^1 \int_0^1 \left(x + y + \frac{5}{3}\right) dy\, dx =$

$\frac{1}{2}\int_0^1 \left(x + \frac{13}{6}\right) dx = \frac{4}{3} \Rightarrow M_{xy} = M_{yz} = M_{xz} = \frac{4}{3}$ by symmetry $\therefore \bar{x} = \bar{y} = \bar{z} = \frac{8}{15}$

$I_z = \int_0^1 \int_0^1 \int_0^1 (x + y + z + 1)(x^2 + y^2)\, dz\, dy\, dx = \int_0^1 \int_0^1 \left(x + y + \frac{3}{2}\right)(x^2 + y^2)\, dy\, dx =$

$\int_0^1 \left(x^3 + 2x^2 + \frac{1}{3}x + \frac{3}{4}\right) dx = \frac{11}{6} \Rightarrow I_x = I_y = I_z = \frac{11}{6}$ by symmetry $\Rightarrow R_x = R_y = R_z = \sqrt{\frac{I_z}{M}} = \sqrt{\frac{11}{15}}$

14.6 TRIPLE INTEGRALS IN CYLINDRICAL AND SPHERICAL COORDINATES

1. $\int_0^{2\pi} \int_0^1 \int_r^{\sqrt{2-r^2}} r\, dz\, dr\, d\theta = \int_0^{2\pi} \int_0^1 \left(\left(2 - r^2\right)^{1/2} r - r^2\right) dr\, d\theta = \int_0^{2\pi} \left(\frac{2^{3/2}}{3} - \frac{2}{3}\right) d\theta = \frac{4\pi\left(\sqrt{2} - 1\right)}{3}$

5. $\int_0^{2\pi} \int_0^1 \int_r^{\left(2-r^2\right)^{-1/2}} 3\, r\, dz\, dr\, d\theta = 3\int_0^{2\pi} \int_0^1 \left(\left(2 - r^2\right)^{-1/2} r - r^2\right) dr\, d\theta = 3\int_0^{2\pi} \left(\sqrt{2} - \frac{4}{3}\right) d\theta = \pi\left(6\sqrt{2} - 8\right)$

9. $\int_0^{2\pi} \int_0^{\pi} \int_0^{(1-\cos\phi)/2} \rho^2 \sin\phi\, d\rho\, d\phi\, d\theta = \frac{1}{24}\int_0^{2\pi} \int_0^{\pi} \left(1 - \cos\phi\right)^3 \sin\phi\, d\phi\, d\theta = \frac{1}{6}\int_0^{2\pi} d\theta = \frac{\pi}{3}$

13. $\int_0^{2\pi} \int_0^3 \int_0^{z/3} r^3\, dr\, dz\, d\theta = \int_0^{2\pi} \int_0^3 \frac{z^4}{324}\, dz\, d\theta = \int_0^{2\pi} \frac{3}{20}\, d\theta = \frac{3\pi}{10}$

17. $\int_0^2 \int_{-\pi}^0 \int_{\pi/4}^{\pi/2} \rho^3 \sin 2\phi\, d\phi\, d\theta\, d\rho = \int_0^2 \int_{-\pi}^0 \frac{\rho^3}{2}\, d\theta\, d\rho = \int_0^2 \frac{\rho^3\pi}{2}\, d\rho = \left[\frac{\pi\rho^4}{8}\right]_0^2 = 2\pi$

21. a) $8\int_0^{\pi/2} \int_0^{\pi/2} \int_0^2 \rho^2 \sin\phi\, d\rho\, d\phi\, d\theta$ b) $8\int_0^{\pi/2} \int_0^2 \int_0^{\sqrt{4-r^2}} r\, dz\, dr\, d\theta$

c) $8\int_0^2 \int_0^{\sqrt{4-x^2}} \int_0^{\sqrt{4-x^2-y^2}} dz\, dy\, dx$

25. $\int_0^{\pi} \int_0^{2\sin\theta} \int_0^{4 - r\sin\theta} f(r,\theta,z)\, dz\, r\, dr\, d\theta$ 29. $\int_0^{\pi/4} \int_0^{\sec\theta} \int_0^{2 - r\sin\theta} f(r,\theta,z)\, dz\, r\, dr\, d\theta$

33. $V = \int_{-\pi/2}^0 \int_0^{3\cos\theta} \int_0^{-r\sin\theta} dz\, r\, dr\, d\theta = \int_{-\pi/2}^0 \int_0^{3\cos\theta} (-r\sin\theta)\, r\, dr\, d\theta = 9\int_{-\pi/2}^0 \cos^3\theta \sin\theta\, d\theta = \frac{9}{4}$

37. $V = 4\int_0^{\pi/2} \int_0^1 \int_0^{r^2} r\, dz\, dr\, d\theta = 4\int_0^{\pi/2} \int_0^1 r^3\, dr\, d\theta = \int_0^{\pi/2} d\theta = \frac{\pi}{2}$

41. $V = 4\int_0^{\pi/2}\int_0^1\int_{4r^2}^{5-r^2} r\,dz\,dr\,d\theta = 4\int_0^{\pi/2}\int_0^1 \left(5-5r^2\right) r\,dr\,d\theta = 5\int_0^{\pi/2} d\theta = \frac{5\pi}{2}$

45. $\text{average} = \frac{1}{2\pi}\int_0^{2\pi}\int_0^1\int_{-1}^1 r^2\,dz\,dr\,d\theta = \frac{1}{2\pi}\int_0^{2\pi}\int_0^1 2r^2\,dr\,d\theta = \frac{1}{3\pi}\int_0^{2\pi} d\theta = \frac{2}{3}$

49. $M = \int_0^{2\pi}\int_0^4\int_0^{\sqrt{r}} dz\,r\,dr\,d\theta = \int_0^{2\pi}\int_0^4 r^{3/2}dr\,d\theta = \frac{64}{5}\int_0^{2\pi} d\theta = \frac{128\pi}{5}$, $M_{xy} = \int_0^{2\pi}\int_0^4\int_0^{\sqrt{r}} z\,dz\,r\,dr\,d\theta =$
$\frac{1}{2}\int_0^{2\pi}\int_0^4 r^2\,dr\,d\theta = \frac{32}{3}\int_0^{2\pi} d\theta = \frac{64\pi}{3}$, $\overline{z} = \frac{M_{xy}}{M} = \frac{5}{6}$, $\overline{x} = \overline{y} = 0$, by symmetry

53. $I_z = \int_0^{2\pi}\int_0^a\int_{-\sqrt{a^2-r^2}}^{\sqrt{a^2-r^2}} r^3\,dz\,dr\,d\theta = -\int_0^{2\pi}\int_0^a r^2\left(a^2-r^2\right)^{1/2}(-2r)\,dr\,d\theta = \int_0^{2\pi}\frac{4a^5}{15}d\theta = \frac{8\pi a^5}{15}$

57. $V = 4\int_0^{\pi/2}\int_0^{\pi}\int_0^{1-\cos\phi} \rho^2\sin\phi\,d\rho\,d\phi\,d\theta = \frac{4}{3}\int_0^{\pi/2}\int_0^{\pi}\left(1-\cos\phi\right)^3\sin\phi\,d\phi\,d\theta = \frac{16}{3}\int_0^{\pi/2} d\theta = \frac{8\pi}{3}$

61. $\int_0^{2\pi}\int_{\pi/3}^{2\pi/3}\frac{a^3}{3}\sin\phi\,d\phi\,d\theta = \frac{a^3}{3}\int_0^{2\pi}d\theta = \frac{2\pi a^3}{3} \Rightarrow V = \frac{4}{3}\pi a^3 - \frac{2\pi a^3}{3} = \frac{2\pi a^3}{3}$

65. $V = 8\int_0^{\pi/2}\int_1^{\sqrt{2}}\int_0^r r\,dz\,dr\,d\theta = 8\int_0^{\pi/2}\int_1^{\sqrt{2}} r^2\,dr\,d\theta = \frac{2\sqrt{2}-1}{3}\int_0^{\pi/2}d\theta = \frac{4\pi(2\sqrt{2}-1)}{3}$

69. $M = \int_0^{2\pi}\int_0^{\pi/4}\int_0^a \rho^2\sin\phi\,d\rho\,d\phi\,d\theta = \frac{a^3}{3}\int_0^{2\pi}\int_0^{\pi/4}\sin\phi\,d\phi\,d\theta = \frac{a^3}{3}\int_0^{2\pi}\frac{\sqrt{2}-1}{\sqrt{2}}d\theta = \frac{\pi a^3\left(2-\sqrt{2}\right)}{3}$

$M_{xy} = \int_0^{2\pi}\int_0^{\pi/4}\int_0^a \rho^3\sin\phi\cos\phi\,d\rho\,d\phi\,d\theta = \frac{a^3}{4}\int_0^{2\pi}\int_0^{\pi/4}\sin\phi\cos\phi\,d\phi\,d\theta = \frac{a^4}{16}\int_0^{2\pi}d\theta = \frac{\pi a^4}{8} \Rightarrow$

$\overline{z} = \frac{3\left(2+\sqrt{2}\right)a}{16}$, $\overline{x} = \overline{y} = 0$ by symmetry

14.7 SUBSTITUTIONS IN MULTIPLE INTEGRALS

1. $\int_0^4\int_{y/2}^{1+y/2}\frac{2x-y}{2}dx\,dy = \int_0^4\left[\frac{x^2}{2}-\frac{xy}{2}\right]_{y/2}^{1+y/2}dy = \frac{1}{2}\int_0^4 dy = 2$

5. $J(u,v) = \begin{vmatrix} v^{-1} & -uv^{-2} \\ v & u \end{vmatrix} = v^{-1}u + v^{-1}u = \frac{2u}{v}$; $\int_1^3\int_1^2 (v+u)\left(\frac{2u}{v}\right)dv\,du = \int_1^3\left[2u + (\ln 4)\,u^2\right]du = 8 + \frac{26\ln 4}{3}$

9. $\begin{vmatrix} \sin\phi\cos\theta & \rho\cos\phi\cos\theta & -\rho\sin\phi\sin\theta \\ \sin\phi\sin\theta & \rho\cos\phi\sin\theta & \rho\sin\phi\cos\theta \\ \cos\phi & -\rho\sin\phi & 0 \end{vmatrix} = \cos\phi\begin{vmatrix} \rho\cos\phi\cos\theta & -\rho\sin\phi\sin\theta \\ \rho\cos\phi\sin\theta & \rho\sin\phi\cos\theta \end{vmatrix} +$

$\rho\sin\phi\begin{vmatrix} \sin\phi\cos\theta & -\rho\sin\phi\sin\theta \\ \sin\phi\sin\theta & \rho\sin\phi\cos\theta \end{vmatrix} = \rho^2\cos\phi\left(\sin\phi\cos\phi\cos^2\theta + \sin\phi\cos\phi\sin^2\theta\right) +$

$\rho^2\sin\phi\left(\sin^2\phi\cos^2\theta + \sin^2\phi\sin^2\theta\right) = \rho^2\sin\phi\cos^2\phi + \rho^2\sin^3\phi = \rho^2\sin\phi\left(\cos^2\phi + \sin^2\phi\right) = \rho^2\sin\phi$

13. $J(u,v,w) = \begin{vmatrix} 1 & 0 & 0 \\ -v/u^2 & 1/u & 0 \\ 0 & 0 & 1/3 \end{vmatrix} = \frac{1}{3u}$; $\int\int_R\int \left(x^2y + 3xyz\right) dx\,dy\,dz =$

$\int\int_G\int \left(u^2\left(\frac{v}{u}\right) + 3u\frac{v}{u}\frac{w}{3}\right) J(u,v,w)\,du\,dv\,dw = \frac{1}{3}\int_0^3\int_0^2\int_1^2 \left(v + \frac{vw}{u}\right) du\,dv\,dw =$

$\frac{1}{3}\int_0^3\int_0^2 (v + vw\ln 2)\,dv\,dw = \frac{1}{3}\int_0^3 \left(2 + (\ln 4)w\right) dw = 2 + \ln 8$

PRACTICE EXERCISES

1. $\int_1^{10}\int_0^{1/y} ye^{xy}\,dx\,dy = \int_1^{10}\left[e^{xy}\right]_0^{1/y} dy =$

$\int_1^{10} (e-1)\,dy = 9e - 9$

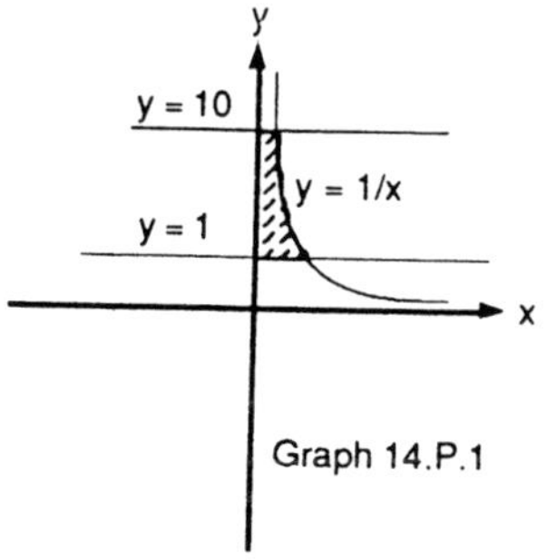

Graph 14.P.1

5. a) $\int_0^{3/2}\int_{-\sqrt{9-4y^2}}^{\sqrt{9-4y^2}} y\,dx\,dy = \int_0^{3/2}\left[\,yx\,\right]_{-\sqrt{9-4y^2}}^{\sqrt{9-4y^2}} dy =$

$\int_0^{3/2} 2y\sqrt{9-4y^2}\,dy = \frac{9}{2}$

b) $\int_{-3}^{3}\int_0^{(9-x^2)^{1/2}/2} y\,dy\,dx = \frac{1}{2}\int_{-3}^{3}\left[y^2\right]_0^{(9-x^2)^{1/2}/2} dx = \frac{1}{8}\int_{-3}^{3}\left(9-x^2\right) dx = \frac{9}{2}$

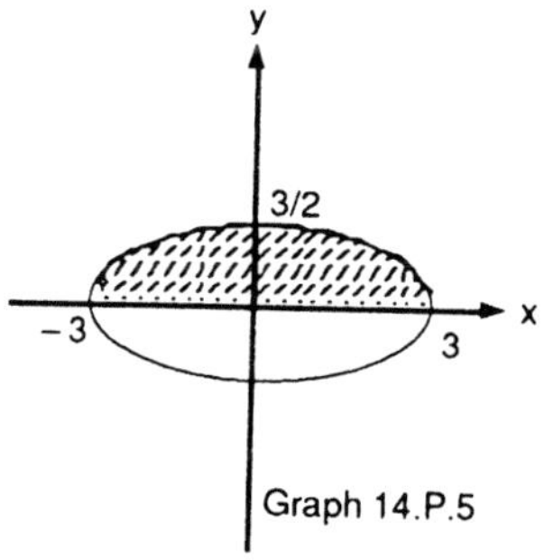

Graph 14.P.5

9. $\int_0^8\int_{\sqrt[3]{x}}^{2} \frac{1}{y^4+1}\,dy\,dx = \int_0^2\int_0^{y^3} \frac{1}{y^4+1}\,dx\,dy = \frac{1}{4}\int_0^2 \frac{4y^3}{y^4+1}\,dy = \frac{\ln 17}{4}$

13. $A = \int_{-2}^{0}\int_{2x+4}^{4-x^2} dy\,dx = \int_{-2}^{0}\left(-x^2 - 2x\right) dx = \frac{4}{3}$

17. $M = \int_1^2\int_{2/x}^{2} dy\,dx = \int_1^2\left(2 - \frac{2}{x}\right) dx = 2 - \ln 4$, $M_y = \int_1^2\int_{2/x}^{2} x\,dy\,dx = \int_1^2 x\left[2 - \frac{2}{x}\right] dx = 1$, $M_x =$

$\int_1^2\int_{2/x}^{2} y\,dy\,dx = \int_1^2\left(2 - \frac{2}{x^2}\right) dx = 1 \Rightarrow \overline{x} = \frac{1}{2 - \ln 4}$, $\overline{y} = \frac{1}{2 - \ln 4}$

21. $M = \int_{-1}^{1}\int_{-1}^{1}\left(x^2 + y^2 + \frac{1}{3}\right) dy\,dx = \int_{-1}^{1}\left(2x^2 + \frac{4}{3}\right) dx = 4$, $M_x = \int_{-1}^{1}\int_{-1}^{1} y\left(x^2 + y^2 + \frac{1}{3}\right) dy\,dx = \int_{-1}^{1} 0\,dx = 0$

$M_y = \int_{-1}^{1}\int_{-1}^{1} x\left(x^2 + y^2 + \frac{1}{3}\right) dy\,dx = \int_{-1}^{1} x\left(2x^2 + \frac{4}{3}\right) dx = 0$

25. $M = 2\int_0^{\pi/2}\int_1^{1+\cos\theta} r\,dr\,d\theta = \int_0^{\pi/2}\left(2\cos\theta + \frac{1+\cos 2\theta}{2}\right)d\theta = \frac{8+\pi}{4}$, $M_y = \int_{-\pi/2}^{\pi/2}\int_1^{1+\cos\theta} \cos\theta\, r^2\,dr\,d\theta =$
$\int_{-\pi/2}^{\pi/2}\left(\cos^2\theta + \cos^3\theta + \frac{\cos^4\theta}{3}\right)d\theta = \frac{32+15\pi}{24} \Rightarrow \overline{x} = \frac{15\pi+32}{6\pi+48}$, $\overline{y} = 0$ by symmetry

29. $\int_0^{\pi}\int_0^{\pi}\int_0^{\pi} \cos(x+y+z)\,dx\,dy\,dz = \int_0^{\pi}\int_0^{\pi}\left[\sin(z+y+\pi) - \sin(z+y)\right]dy\,dz =$
$\int_0^{\pi}\left(-\cos(z+2\pi) + \cos(z+\pi) + \cos(z) - \cos(z+\pi)\right)dz = 0$

33. $V = 2\int_0^{\pi/2}\int_{-\cos y}^{0}\int_0^{-2x} dz\,dx\,dy = -4\int_0^{\pi/2}\int_{-\cos y}^{0} x\,dx\,dy = 2\int_0^{\pi/2}\cos^2 y\,dy = \frac{\pi}{2}$

37. a) $\int_{-\sqrt{2}}^{\sqrt{2}}\int_{-\sqrt{2-y^2}}^{\sqrt{2-y^2}}\int_{\sqrt{x^2+y^2}}^{\sqrt{4-(x^2+y^2)}} 3\,dz\,dx\,dy$

b) $\int_0^{2\pi}\int_0^{\pi/4}\int_0^{2} 3\rho^2\sin\phi\,d\rho\,d\phi\,d\theta$

c) $\int_0^{2\pi}\int_0^{\sqrt{2}}\int_r^{\sqrt{4-r^2}} 3\,dz\,r\,dr\,d\theta = 3\int_0^{2\pi}\int_0^{\sqrt{2}}\left(\left(4-r^2\right)^{1/2} - r\right)r\,dr\,d\theta = \int_0^{2\pi}\left(8 - 2^{5/2}\right)d\theta = 2\pi\left(8 - 2^{5/2}\right)$

41. $\int_0^{1}\int_{\sqrt{1-x^2}}^{\sqrt{3-x^2}}\int_1^{\sqrt{4-x^2-y^2}} yxz^2\,dz\,dy\,dx + \int_1^{\sqrt{3}}\int_0^{\sqrt{3-x^2}}\int_1^{\sqrt{4-x^2-y^2}} yxz^2\,dz\,dy\,dx$

45. $V = \int_0^{\pi/2}\int_1^{2}\int_0^{r^2\sin\theta\cos\theta} r\,dz\,dr\,d\theta = \int_0^{\pi/2}\int_1^{2} r^3\sin\theta\cos\theta\,dr\,d\theta = \frac{15}{4}\int_0^{\pi/2}\sin\theta\cos\theta\,d\theta = \frac{15}{8}$

49. a) $M = 4\int_0^{\pi/2}\int_0^{1}\int_0^{r^2} z\,r\,dz\,dr\,d\theta = 2\int_0^{\pi/2}\int_0^{1} r^5\,dr\,d\theta = \frac{1}{3}\int_0^{\pi/2} d\theta = \frac{\pi}{6}$

$M_{xy} = \int_0^{2\pi}\int_0^{1}\int_0^{r^2} z^2\,r\,dz\,dr\,d\theta = \frac{1}{3}\int_0^{2\pi}\int_0^{1} r^7\,dr\,d\theta = \frac{1}{24}\int_0^{2\pi} d\theta = \frac{\pi}{12} \Rightarrow \overline{z} = \frac{1}{2}$, $\overline{x} = \overline{y} = 0$ by symmetry;

$I_z = \int_0^{2\pi}\int_0^{1}\int_0^{r^2} z\,r^3\,dz\,dr\,d\theta = \frac{1}{2}\int_0^{2\pi}\int_0^{1} r^7\,dr\,d\theta = \frac{1}{16}\int_0^{2\pi} d\theta = \frac{\pi}{8} \Rightarrow R_z = \sqrt{\frac{I_z}{M}} = \frac{\sqrt{3}}{2}$

b) $M = 4\int_0^{\pi/2}\int_0^{1}\int_0^{r^2} r^2\,dz\,dr\,d\theta = 4\int_0^{\pi/2}\int_0^{1} r^4\,dr\,d\theta = \frac{4}{5}\int_0^{\pi/2} d\theta = \frac{2\pi}{5}$

$M_{xy} = \int_0^{2\pi}\int_0^{1}\int_0^{r^2} z\,r^2\,dz\,dr\,d\theta = \frac{1}{2}\int_0^{2\pi}\int_0^{1} r^6\,dr\,d\theta = \frac{1}{14}\int_0^{2\pi} d\theta = \frac{\pi}{7} \Rightarrow \overline{z} = \frac{5}{14}$, $\overline{x} = \overline{y} = 0$ by symmetry;

$I_z = \int_0^{2\pi}\int_0^{1}\int_0^{r^2} r^4\,dz\,dr\,d\theta = \int_0^{2\pi}\int_0^{1} r^6\,dr\,d\theta = \frac{1}{7}\int_0^{2\pi} d\theta = \frac{2\pi}{7} \Rightarrow R_z = \sqrt{\frac{I_z}{M}} = \sqrt{\frac{5}{7}}$

53. $M = \frac{2}{3}\pi a^3$; $M_{xy} = \int_0^{2\pi}\int_0^{\pi/2}\int_0^{a} \rho^3\sin\phi\cos\phi\,d\rho\,d\phi\,d\theta = \frac{a^4}{4}\int_0^{2\pi}\int_0^{\pi/2}\sin\phi\cos\phi\,d\phi\,d\theta =$
$\frac{a^4}{8}\int_0^{2\pi} d\theta = \frac{a^4\pi}{4} \Rightarrow \overline{z} = \frac{3a}{8}$, $\overline{x} = \overline{y} = 0$ by symmetry

CHAPTER 15

INTEGRATION IN VECTOR FIELDS

15.1 LINE INTEGRALS

1. $\mathbf{r} = t\,\mathbf{i} + (1-t)\,\mathbf{j} \Rightarrow x = t,\ y = 1 - t \Rightarrow y = 1 - x \Rightarrow$ c

5. $\mathbf{r} = t\,\mathbf{i} + t\,\mathbf{j} + t\,\mathbf{k} \Rightarrow x = t,\ y = t,\ z = t \Rightarrow$ d

9. $\mathbf{r} = t\,\mathbf{i} + (1-t)\,\mathbf{j} \Rightarrow x = t,\ y = 1 - t,\ z = 0 \Rightarrow f(g(t),h(t),k(t)) = 1.\ \frac{dx}{dt} = 1, \frac{dy}{dt} = -1, \frac{dz}{dt} = 0 \Rightarrow$

$$\sqrt{\left(\frac{dx}{dt}\right)^2 + \left(\frac{dy}{dt}\right)^2 + \left(\frac{dz}{dt}\right)^2}\ dt = \sqrt{2}\,dt \Rightarrow \int_C f(x,y,z)\,ds = \int_0^1 \sqrt{2}\,dt = \sqrt{2}$$

13. $\mathbf{r} = \mathbf{i} + \mathbf{j} + t\,\mathbf{k} \Rightarrow x = 1,\ y = 1,\ z = t \Rightarrow f(g(t),h(t),k(t)) = 3t\sqrt{4 + t^2}\quad \frac{dx}{dt} = 0, \frac{dy}{dt} = 0, \frac{dz}{dt} = 1 \Rightarrow$

$$\sqrt{\left(\frac{dx}{dt}\right)^2 + \left(\frac{dy}{dt}\right)^2 + \left(\frac{dz}{dt}\right)^2}\ dt = 1\,dt = dt \Rightarrow \int_C f(x,y,z)\,ds = \int_{-1}^1 3t\sqrt{4 + t^2}\ dt = 0$$

17. $\mathbf{r} = t\,\mathbf{i} + t\,\mathbf{j} + t\,\mathbf{k} \Rightarrow x = t,\ y = t,\ z = t \Rightarrow f(g(t),h(t),k(t)) = \frac{t + t + t}{t^2 + t^2 + t^2} = \frac{1}{t}.\ \frac{dx}{dt} = 1, \frac{dy}{dt} = 1, \frac{dz}{dt} = 1 \Rightarrow$

$$\sqrt{\left(\frac{dx}{dt}\right)^2 + \left(\frac{dy}{dt}\right)^2 + \left(\frac{dz}{dt}\right)^2}\ dt = \sqrt{3}\,dt \Rightarrow \int_C f(x,y,z)\,ds = \int_a^b \frac{1}{t}\left(\sqrt{3}\ dt\right) = \sqrt{3}\ln|b| - \sqrt{3}\ln|a|$$

$$= \sqrt{3}\ln\left|\frac{b}{a}\right|$$

21. Let δ be constant. Let $x = a\cos t$, $y = a\sin t$. Then $\frac{dx}{dt} = -a\sin t, \frac{dy}{dt} = a\cos t, 0 \le t \le 2\pi, \frac{dz}{dt} = 0 \Rightarrow$

$$\sqrt{\left(\frac{dx}{dt}\right)^2 + \left(\frac{dy}{dt}\right)^2 + \left(\frac{dz}{dt}\right)^2}\ dt = a\,dt.\quad \therefore\ I_z = \int_C (x^2 + y^2)\delta\,ds = \int_0^{2\pi} (a^2\sin^2 t + a^2\cos^2 t)a\delta\,dt =$$

$$\int_0^{2\pi} a^3\delta\,dt = 2\pi a^3\delta.\quad M = \int_C \delta(x,y,z)\,ds = \int_0^{2\pi} \delta a\,dt = 2\pi\delta a.\quad R_z = \sqrt{\frac{I_z}{M}} = \sqrt{\frac{2\pi a^3\delta}{2\pi a\delta}} = a.$$

25. $\mathbf{r} = (1-t)\mathbf{i} + (1-t)\mathbf{j} + (1-t)\mathbf{k},\ 0 \le t \le 1 \Rightarrow \mathbf{v}(t) = -\mathbf{i} - \mathbf{j} - \mathbf{k} \Rightarrow |\mathbf{v}(t)| = \sqrt{(-1)^2 + (-1)^2 + (-1)^2}$

$= \sqrt{3}$. The integral of f over C is $\int_C f(x,y,z)\,ds = \int_0^1 f(1 - t,\ 1 - t,\ 1 - t)\left(\sqrt{3}\right)dt =$

$$\int_0^1 \left((1 - t) - 3(1 - t)^2 + (1 - t)\right)\sqrt{3}\ dt = \sqrt{3}\int_0^1 \left(-1 + 4t - 3t^2\right)dt = \sqrt{3}\left[-t + 2t^2 - t^3\right]_0^1 = 0$$

15.2 VECTOR FIELDS, WORK, CIRCULATION, AND FLUX

1. $f(x,y,z) = \left(x^2 + y^2 + z^2\right)^{-1/2} \Rightarrow \frac{\partial f}{\partial x} = -\frac{1}{2}\left(x^2 + y^2 + z^2\right)^{-3/2}(2x) = -x\left(x^2 + y^2 + z^2\right)^{-3/2}$. Similarly, $\frac{\partial f}{\partial y} = -y\left(x^2 + y^2 + z^2\right)^{-3/2}, \frac{\partial f}{\partial z} = -z\left(x^2 + y^2 + z^2\right)^{-3/2}$. $\therefore \nabla f = \frac{-x\,\mathbf{i} - y\,\mathbf{j} - z\,\mathbf{k}}{\left(x^2 + y^2 + z^2\right)^{3/2}}$.

5. a) $\mathbf{F} = 3t\,\mathbf{i} + 2t\,\mathbf{j} + 4t\,\mathbf{k}, \frac{d\mathbf{r}}{dt} = \mathbf{i} + \mathbf{j} + \mathbf{k} \Rightarrow \mathbf{F} \cdot \frac{d\mathbf{r}}{dt} = 9t \Rightarrow W = \int_0^1 9t\,dt = \frac{9}{2}$

b) $\mathbf{F} = 3t^2\,\mathbf{i} + 2t\,\mathbf{j} + 4t^4\,\mathbf{k}, \frac{d\mathbf{r}}{dt} = \mathbf{i} + 2t\,\mathbf{j} + 4t^3\,\mathbf{k} \Rightarrow \mathbf{F} \cdot \frac{d\mathbf{r}}{dt} = 7t^2 + 16t^7 \Rightarrow W = \int_0^1 (7t^2 + 16t^7)\,dt = \frac{13}{3}$

c) $\mathbf{F}_1 = 3t\,\mathbf{i} + 2t\,\mathbf{j}, \frac{d\mathbf{r}_1}{dt} = \mathbf{i} + \mathbf{j} \Rightarrow \mathbf{F}_1 \cdot \frac{d\mathbf{r}_1}{dt} = 5t \Rightarrow W_1 = \int_0^1 5t\,dt = \frac{5}{2}$. $\mathbf{F}_2 = 3\,\mathbf{i} + 2\,\mathbf{j} + 4t\,\mathbf{k}, \frac{d\mathbf{r}_2}{dt} = \mathbf{k} \Rightarrow$

$\mathbf{F}_2 \cdot \frac{d\mathbf{r}_2}{dt} = 4t \Rightarrow W_2 = \int_0^1 4t\,dt = 2$. $\therefore\ W = W_1 + W_2 = \frac{9}{2}$

9. a) $\mathbf{F} = (3t^2 - 3t)\,\mathbf{i} + 3t\,\mathbf{j} + \mathbf{k}, \frac{d\mathbf{r}}{dt} = \mathbf{i} + \mathbf{j} + \mathbf{k} \Rightarrow \mathbf{F} \cdot \frac{d\mathbf{r}}{dt} = 3t^2 + 1 \Rightarrow W = \int_0^1 (3t^2 + 1)\,dt = 2$

b) $\mathbf{F} = (3t^2 - 3t)\,\mathbf{i} + 3t^4\,\mathbf{j} + \mathbf{k}, \frac{d\mathbf{r}}{dt} = \mathbf{i} + 2t\,\mathbf{j} + 4t^3\,\mathbf{k} \Rightarrow \mathbf{F} \cdot \frac{d\mathbf{r}}{dt} = 6t^5 + 4t^3 + 3t^2 - 3t \Rightarrow$

$W = \int_0^1 \left(6t^5 + 4t^3 + 3t^2 - 3t\right) dt = \frac{3}{2}$

c) $\mathbf{F}_1 = (3t^2 - 3t)\,\mathbf{i} + \mathbf{k}, \frac{d\mathbf{r}_1}{dt} = \mathbf{i} + \mathbf{j} \Rightarrow \mathbf{F}_1 \cdot \frac{d\mathbf{r}_1}{dt} = 3t^2 - 3t \Rightarrow W_1 = \int_0^1 (3t^2 - 3t)\,dt = -\frac{1}{2}$

$\mathbf{F}_2 = 3t\,\mathbf{j} + \mathbf{k}, \frac{d\mathbf{r}_2}{dt} = \mathbf{k} \Rightarrow \mathbf{F}_2 \cdot \frac{d\mathbf{r}_2}{dt} = 1 \Rightarrow W_2 = \int_0^1 dt = 1$. $\therefore\ W = W_1 + W_2 = \frac{1}{2}$

13. $\mathbf{F} = t\,\mathbf{i} + (\sin t)\,\mathbf{j} + (\cos t)\,\mathbf{k}, \frac{d\mathbf{r}}{dt} = (\cos t)\,\mathbf{i} - (\sin t)\,\mathbf{j} + \mathbf{k} \Rightarrow \mathbf{F} \cdot \frac{d\mathbf{r}}{dt} = t\cos t - \sin^2 t + \cos t \Rightarrow$

$W = \int_0^{2\pi} (t\cos t - \sin^2 t + \cos t)\,dt = -\pi$

17. $\mathbf{F} = (\cos t - \sin t)\,\mathbf{i} + (\cos t)\,\mathbf{k}, \frac{d\mathbf{r}}{dt} = (-\sin t)\,\mathbf{i} + (\cos t)\,\mathbf{k} \Rightarrow \mathbf{F} \cdot \frac{d\mathbf{r}}{dt} = -\sin t\cos t + 1 \Rightarrow$

$\text{Flow} = \int_0^{\pi} (-\sin t\cos t + 1)\,dt = \pi$

21. $\mathbf{F}_1 = (a\cos t)\,\mathbf{i} + (a\sin t)\,\mathbf{j},\ \frac{d\mathbf{r}_1}{dt} = (-a\sin t)\,\mathbf{i} + (a\cos t)\,\mathbf{j} \Rightarrow \mathbf{F}_1 \cdot \frac{d\mathbf{r}_1}{dt} = 0 \Rightarrow \text{Circ}_1 = 0.$ $M_1 = a\cos t,\ N_1 =$ $a\sin t,\ dx = -a\sin t\,dt,\ dy = a\cos t\,dt \Rightarrow \text{Flux}_1 = \int_C M_1 dy - N_1 dx = \int_0^\pi (a^2\cos^2 t + a^2\sin^2 t)\,dt =$ $\int_0^\pi a^2\,dt = a^2\pi.$

$\mathbf{F}_2 = t\,\mathbf{i},\ \frac{d\mathbf{r}_2}{dt} = \mathbf{i} \Rightarrow \mathbf{F}_2 \cdot \frac{d\mathbf{r}_2}{dt} = t \Rightarrow \text{Circ}_2 = \int_{-a}^{a} t\,dt = 0.$ $M_2 = t,\ N_2 = 0,\ dx = dt,\ dy = 0 \Rightarrow \text{Flux}_2 =$ $\int_C M_2 dy - N_2 dx = \int_{-a}^{a} 0\,dt = 0.$ $\therefore$ Circ = $\text{Circ}_1 + \text{Circ}_2 = 0$, Flux = $\text{Flux}_1 + \text{Flux}_2 = a^2\pi$

25.

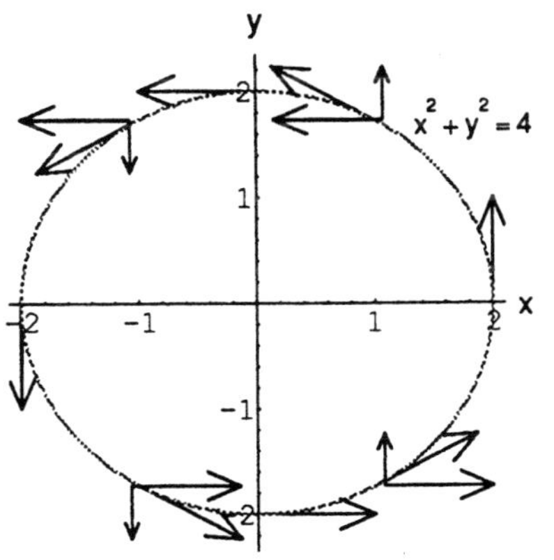

Graph 15.2.25

$\mathbf{F} = -\frac{y}{\sqrt{x^2+y^2}}\,\mathbf{i} + \frac{x}{\sqrt{x^2+y^2}}\,\mathbf{j}$ on $x^2 + y^2 = 4$. At (2,0), $\mathbf{F} = \mathbf{j}$, at (0,2), $\mathbf{F} = -\mathbf{i}$, at (−2,0), $\mathbf{F} = -\mathbf{j}$, at (0,−2), $\mathbf{F} = \mathbf{i}$. At $(1,\sqrt{3})$, $\mathbf{F} = -\frac{\sqrt{3}}{2}\mathbf{i} + \frac{1}{2}\mathbf{j}$, at $(1,-\sqrt{3})$, $\mathbf{F} = \frac{\sqrt{3}}{2}\mathbf{i} + \frac{1}{2}\mathbf{j}$, at $(-1,\sqrt{3})$, $\mathbf{F} = -\frac{\sqrt{3}}{2}\mathbf{i} - \frac{1}{2}\mathbf{j}$, at $(-1,-\sqrt{3})$, $\mathbf{F} = \frac{\sqrt{3}}{2}\mathbf{i} - \frac{1}{2}\mathbf{j}$

29. The slope of a line through (x,y) and the origin is $\frac{y}{x} \Rightarrow \mathbf{v} = x\,\mathbf{i} + y\,\mathbf{j}$ is a vector on that line. But $\mathbf{v}$ points away from the origin. $\therefore\ \mathbf{F} = -\frac{x\,\mathbf{i} + y\,\mathbf{j}}{\sqrt{x^2+y^2}}$ is the unit vector we want.

33. C_1: $\mathbf{r}(t) = \cos t\,\mathbf{i} + \sin t\,\mathbf{j} + t\,\mathbf{k},\ 0 \le t \le \frac{\pi}{2} \Rightarrow \mathbf{F} = 2\cos t\,\mathbf{i} + 2t\,\mathbf{j} + 2\sin t\,\mathbf{k}.$ $\frac{d\mathbf{r}}{dt} = -\sin t\,\mathbf{i} + \cos t\,\mathbf{j} + \mathbf{k} \Rightarrow \mathbf{F} \cdot \frac{d\mathbf{r}}{dt} =$ $-2\cos t\sin t + 2t\cos t + 2\sin t = -\sin 2t + 2t\cos t + 2\sin t$. Then $\text{Flow}_1 =$

$$\int_0^{\pi/2} (-\sin 2t + 2t\cos t + 2\sin t)\,dt = \left[\frac{1}{2}\cos 2t + 2t\sin t + 2\cos t - 2\cos t\right]_0^{\pi/2}$$

(do the middle term of the integral by Parts) $= -1 + \pi$.

C_2: $\mathbf{r}(t) = \mathbf{j} + \frac{\pi}{2}(1-t)\,\mathbf{k},\ 0 \le t \le 1 \Rightarrow \mathbf{F} = \pi(1-t)\,\mathbf{j} + 2\,\mathbf{k}.$ $\frac{d\mathbf{r}}{dt} = -\frac{\pi}{2}\mathbf{k} \Rightarrow \mathbf{F} \cdot \frac{d\mathbf{r}}{dt} = -\pi.$ $\therefore\ \text{Flow}_2 =$ $\int_0^1 -\pi\,dt = \left[-\pi t\right]_0^1 = -\pi.$

C_3: $\mathbf{r}(t) = t\,\mathbf{i} + (1-t)\,\mathbf{j},\ 0 \le t \le 1 \Rightarrow \mathbf{F} = 2t\,\mathbf{i} + 2(1-t)\,\mathbf{k}.$ $\frac{d\mathbf{r}}{dt} = \mathbf{i} - \mathbf{j} \Rightarrow \mathbf{F} \cdot \frac{d\mathbf{r}}{dt} = 2t.$ $\therefore\ \text{Flow}_3 =$ $\int_0^1 2t\,dt = \left[t^2\right]_0^1 = 1.$ $\therefore$ Circulation $= (-1 + \pi) - \pi + 1 = 0.$

15.3 PATH INDEPENDENCE, POTENTIAL FUNCTIONS, AND CONSERVATIVE FIELDS

1. $\frac{\partial P}{\partial y} = x = \frac{\partial N}{\partial z}, \frac{\partial M}{\partial z} = y = \frac{\partial P}{\partial x}, \frac{\partial N}{\partial x} = z = \frac{\partial M}{\partial y} \Rightarrow$ Conservative

5. $\frac{\partial N}{\partial x} = 0 \neq \frac{\partial M}{\partial y} \Rightarrow$ Not Conservative

9. $\frac{\partial f}{\partial x} = e^{y+2z} \Rightarrow f(x,y,z) = x\,e^{y+2z} + g(y,z).\ \frac{\partial f}{\partial y} = x\,e^{y+2z} + \frac{\partial g}{\partial y} = x\,e^{y+2z} \Rightarrow \frac{\partial g}{\partial y} = 0.$ Then $f(x,y,z) = x\,e^{y+2z} + h(z).\ \frac{\partial f}{\partial z} = 2x\,e^{y+2z} + h'(z) = 2x\,e^{y+2z} \Rightarrow h'(z) = 0 \Rightarrow h(z) = C.\ \therefore\ f(x,y,z) = x\,e^{y+2z} + C$

13. Let $F(x,y,z) = 2xy\,\mathbf{i} + (x^2 - z^2)\,\mathbf{j} - 2yz\,\mathbf{k} \Rightarrow \frac{\partial P}{\partial y} = -2z = \frac{\partial N}{\partial z}, \frac{\partial M}{\partial z} = 0 = \frac{\partial P}{\partial x}, \frac{\partial N}{\partial x} = 2x = \frac{\partial M}{\partial y} \Rightarrow M\,dx + N\,dy + P\,dz$ is exact. $\frac{\partial f}{\partial x} = 2xy \Rightarrow f(x,y,z) = x^2y + g(y,z).\ \frac{\partial f}{\partial y} = x^2 + \frac{\partial g}{\partial y} = x^2 - z^2 \Rightarrow \frac{\partial g}{\partial y} = -z^2 \Rightarrow g(y,z) = -yz^2 + h(z) \Rightarrow f(x,y,z) = x^2y - yz^2 + h(z).\ \frac{\partial f}{\partial z} = -2yz + h'(z) = -2yz \Rightarrow h'(z) = 0 \Rightarrow h(z) = C \Rightarrow f(x,y,z) = x^2y - yz^2 + C \Rightarrow \int_{(0,0,0)}^{(1,2,3)} 2xy\,dx + (x^2 - z^2)\,dy - 2yz\,dz = f(1,2,3) - f(0,0,0) = -16$

17. Let $F(x,y,z) = (2\cos y)\,\mathbf{i} + \left(\frac{1}{y} - 2x\sin y\right)\mathbf{j} + \frac{1}{z}\mathbf{k} \Rightarrow \frac{\partial P}{\partial y} = 0 = \frac{\partial N}{\partial z}, \frac{\partial M}{\partial z} = 0 = \frac{\partial P}{\partial x}, \frac{\partial N}{\partial x} = -2\sin y = \frac{\partial M}{\partial y} \Rightarrow M\,dx + N\,dy + P\,dz$ is exact. $\frac{\partial f}{\partial x} = 2\cos y \Rightarrow f(x,y,z) = 2x\cos y + g(y,z).\ \frac{\partial f}{\partial y} = -2x\sin y + \frac{\partial g}{\partial y} = \frac{1}{y} - 2x\sin y \Rightarrow \frac{\partial g}{\partial y} = \frac{1}{y} \Rightarrow g(y,z) = \ln y + h(z) \Rightarrow f(x,y,z) = 2x\cos y + \ln y + h(z).\ \frac{\partial f}{\partial z} = h'(z) = \frac{1}{z} \Rightarrow h(z) = \ln z + C \Rightarrow f(x,y,z) = 2x\cos y + \ln y + \ln z + C \Rightarrow \int_{(0,2,1)}^{(1,\pi/2,2)} 2\cos y\,dx + \left(\frac{1}{y} - 2x\sin y\right)dy + \frac{1}{z}dz = f(1,\frac{\pi}{2},2) - f(0,2,1) = \ln\frac{\pi}{2}$

21. Let $x - 1 = t,\ y - 1 = 2t,\ z - 1 = -2t,\ 0 \le t \le 1 \Rightarrow dx = dt,\ dy = 2\,dt,\ dz = -2\,dt \Rightarrow \int_{(1,1,1)}^{(2,3,-1)} y\,dx + x\,dy + 4\,dz$

$$= \int_0^1 (2t+1)dt + (t+1)2\,dt + 4(-2)dt = \int_0^1 (4t-5)\,dt = -3$$

25. $\frac{\partial P}{\partial y} = 0 = \frac{\partial N}{\partial z}, \frac{\partial M}{\partial z} = 0 = \frac{\partial P}{\partial x}, \frac{\partial N}{\partial x} = -\frac{2x}{y^2} = \frac{\partial M}{\partial y} \Rightarrow F$ is conservative $\Rightarrow$ there exists an f so that $F = \nabla f$. $\frac{\partial f}{\partial x} = \frac{2x}{y} \Rightarrow f(x,y) = \frac{x^2}{y} + g(y) \Rightarrow \frac{\partial f}{\partial y} = -\frac{x^2}{y} + g'(y) = \frac{1 - x^2}{y^2} \Rightarrow g'(y) = \frac{1}{y^2} \Rightarrow g(y) = -\frac{1}{y} + C \Rightarrow f(x,y) = \frac{x^2}{y} - \frac{1}{y} + C.$ Let $C = 0$. Then $f(x,y) = \frac{x^2 - 1}{y}.\ \therefore\ F = \nabla\left(\frac{x^2 - 1}{y}\right).$

29\. a) $\mathbf{F} = \nabla(x^3y^2) \Rightarrow \mathbf{F} = 3x^2y^2\,\mathbf{i} + 2x^3y\,\mathbf{j}$. Let C_1 be the path from $(-1,1)$ to $(0,0) \Rightarrow x = t-1,\ y = -t+1,$ $0 \le t \le 1 \Rightarrow \mathbf{F} = 3(t-1)^2(-t+1)^2\,\mathbf{i} + 2(t-1)^3(-t+1)\,\mathbf{j} = 3(t-1)^4\,\mathbf{i} - 2(t-1)^4\,\mathbf{j}$ and $\mathbf{R}_1 = (t-1)\,\mathbf{i} + (-t+1)\,\mathbf{j} \Rightarrow d\mathbf{R}_1 = dt\,\mathbf{i} - dt\,\mathbf{j}$. $\therefore \displaystyle\int_C \mathbf{F}\cdot d\mathbf{R}_1 = \int_0^1 \left(3(t-1)^4 + 2(t-1)^4\right)dt = \int_0^1 5(t-1)^4\,dt = 1$

Let C_2 be the path from $(0,0)$ to $(1,1)$ y $x = t,\ y = t,\ 0 \le t \le 1 \Rightarrow \mathbf{F} = 3t^4\,\mathbf{i} + 2t^4\,\mathbf{j}$ and $\mathbf{R}_2 = t\,\mathbf{i} + t\,\mathbf{j} \Rightarrow$ $d\mathbf{R}_2 = dt\,\mathbf{i} + dt\,\mathbf{j}$. $\therefore \displaystyle\int_{C_2} \mathbf{F}\cdot d\mathbf{R}_2 = \int_0^1 (3t^4 + 2t^4)\,dt = \int_0^1 5t^4\,dt = 1$. Then $\displaystyle\int_C \mathbf{F}\cdot d\mathbf{R} =$

$$\int_{C_1} \mathbf{F}\cdot d\mathbf{R}_1 + \int_{C_2} \mathbf{F}\cdot d\mathbf{R}_2 = 2$$

b) Since $f(x,y) = x^3y^2$ is a potential function for $\mathbf{F}$, $\displaystyle\int_{(-1,1)}^{(1,1)} \mathbf{F}\cdot d\mathbf{R} = f(1,1) - f(-1,1) = 2.$

15.4 GREEN'S THEOREM IN THE PLANE

1\. Equation 15: $M = -y = -a\sin t,\ N = x = a\cos t,\ dx = -a\sin t\,dt,\ dy = a\cos t\,dt \Rightarrow \dfrac{\partial M}{\partial x} = 0, \dfrac{\partial M}{\partial y} = -1,$ $\dfrac{\partial N}{\partial x} = 1, \dfrac{\partial N}{\partial y} = 0 \Rightarrow \displaystyle\oint_C M\,dy - N\,dx = \int_0^{2\pi} \left((-a\sin t)(a\cos t)\,dt - (a\cos t)(-a\sin t)\right)dt =$ $0.\ \displaystyle\int_R\int \left(\frac{\partial M}{\partial x} + \frac{\partial N}{\partial y}\right)dx\,dy = \int_R\int 0\,dx\,dy = 0$

Equation 16: $\displaystyle\oint_C M\,dx + N\,dy = \int_0^{2\pi} \left((-a\sin t)(-a\sin t) + (a\cos t)(a\cos t)\right)dt = 2\pi a^2$

$$\int_R\int \left(\frac{\partial N}{\partial x} - \frac{\partial M}{\partial y}\right)dx\,dy = \int_{-a}^{a}\int_{-\sqrt{a^2-x^2}}^{\sqrt{a^2-x^2}} 2\,dy\,dx \quad \int_{-a}^{a} 4\sqrt{a^2-x^2}\,dx = 2a^2\pi$$

5\. $M = x - y,\ N = y - x \Rightarrow \dfrac{\partial M}{\partial x} = 1, \dfrac{\partial M}{\partial y} = -1, \dfrac{\partial N}{\partial x} = -1, \dfrac{\partial N}{\partial y} = 1 \Rightarrow \text{Flux} = \displaystyle\int_R\int 2\,dx\,dy = \int_0^1\int_0^1 2\,dx\,dy$ $= 2$. Circ $= \displaystyle\int_R\int (-1 - (-1))\,dx\,dy = 0$

9. $M = x + e^x \sin y,\ N = x + e^x \cos y \Rightarrow \frac{\partial M}{\partial x} = 1 + e^x \sin y,\ \frac{\partial M}{\partial y} = e^x \cos y,\ \frac{\partial N}{\partial x} = 1 + e^x \cos y,\ \frac{\partial N}{\partial y} = e^x \sin y$

$$\Rightarrow \text{Flux} = \int_R\int dx\,dy = \int_{-\pi/4}^{\pi/4}\int_0^{\sqrt{\cos 2\theta}} r\,dr\,d\theta = \int_{-\pi/4}^{\pi/4}\left(\frac{1}{2}\cos 2\theta\right)d\theta = \frac{1}{2}.$$

$$\text{Circ} = \int_R\int (1 + e^x \cos y - e^x \cos y)\,dx\,dy = \int_R\int dx\,dy =$$

$$\int_{-\pi/4}^{\pi/4}\int_0^{\sqrt{\cos 2\theta}} r\,dr\,d\theta = \int_{-\pi/4}^{\pi/4}\frac{1}{2}\cos 2\theta\,d\theta = \frac{1}{2}$$

13. $M = 3xy - \frac{x}{1+y^2},\ N = e^x + \tan^{-1} y \Rightarrow \frac{\partial M}{\partial x} = 3y - \frac{1}{1+y^2},\ \frac{\partial N}{\partial y} = \frac{1}{1+y^2} \Rightarrow \text{Flux} =$

$$\int_R\int\left(3y - \frac{1}{1+y^2} + \frac{1}{1+y^2}\right)dx\,dy = \int_R\int 3y\,dx\,dy = \int_0^{2\pi}\int_0^{a(1+\cos\theta)} 3r\sin\theta\,r\,dr\,d\theta = \int_0^{2\pi} a^3(1+\cos\theta)^3\sin\theta\,d\theta$$

$= -4a^3 - (-4a^3) = 0$

17. $M = 6y + x,\ N = y + 2x \Rightarrow \frac{\partial M}{\partial y} = 6,\ \frac{\partial N}{\partial x} = 2 \Rightarrow \oint_C (6y + x)dx + (y + 2x)dy = \int_R\int (2 - 6)\,dy\,dx =$

$-4(\text{Area of the circle}) = -16\pi$

21. a) $M = f(x),\ N = g(y) \Rightarrow \frac{\partial M}{\partial y} = 0,\ \frac{\partial N}{\partial x} = 0 \Rightarrow \oint_C f(x)\,dx + g(y)\,dy = \int_R\int 0\,dy\,dx = 0$

b) $M = ky,\ N = hx \Rightarrow \frac{\partial M}{\partial y} = k,\ \frac{\partial N}{\partial x} = h \Rightarrow \oint_C ky\,dx + hx\,dy = \int_R\int (h - k)\,dx\,dy =$

$(h - k)(\text{Area of the region})$

25. If a two-dimensional vector field is conservative, then $\frac{\partial N}{\partial x} = \frac{\partial M}{\partial y} \Rightarrow \frac{\partial N}{\partial x} - \frac{\partial M}{\partial y} = \text{curl } \mathbf{F} = 0$. A two-dimensional field $\mathbf{F} = M\,\mathbf{i} + N\,\mathbf{j}$ can be considered to be the restriction to the xy-plane of a three-dimensional field whose **k** component is zero, and whose **i** and **j** components are independent of z.

29. $\text{Area} = \frac{1}{2}\oint_C x\,dy - y\,dx.\ M = x = \cos^3 t,\ N = y = \sin^3 t \Rightarrow dx = -3\cos^2 t\sin t\,dt,\ dy = 3\sin^2 t\cos t\,dt$

$$\Rightarrow \text{Area} = \frac{1}{2}\int_0^{2\pi}(3\sin^2 t\cos^2 t(\cos^2 t + \sin^2 t))\,dt = \frac{1}{2}\int_0^{2\pi}(3\sin^2 t\cos^2 t)\,dt = \frac{3\pi}{8}$$

15.5 SURFACE AREA AND SURFACE INTEGRALS

1. $\mathbf{p} = \mathbf{k}$, $\nabla f = 2x\,\mathbf{i} + 2y\,\mathbf{j} - \mathbf{k} \Rightarrow |\nabla f| = \sqrt{(2x)^2 + (2y)^2 + (-1)^2} = \sqrt{4x^2 + 4y^2 + 1}$. $|\nabla f \cdot \mathbf{p}| = 1 \Rightarrow$

$$S = \iint_R \frac{|\nabla f|}{|\nabla f \cdot \mathbf{p}|}\,dA = \iint_R \sqrt{4x^2 + 4y^2 + 1}\;dx\,dy =$$

$$\iint_R \sqrt{4r^2\cos^2\theta + 4r^2\sin^2\theta + 1}\;r\,dr\,d\theta = \int_0^{2\pi}\int_0^{\sqrt{2}} \sqrt{4r^2 + 1}\;r\,dr\,d\theta = \frac{13}{3}\pi$$

5. $\mathbf{p} = \mathbf{k}$. $\nabla f = 2x\,\mathbf{i} - 2\,\mathbf{j} - 2\,\mathbf{k} \Rightarrow |\nabla f| = \sqrt{(2x)^2 + (-2)^2 + (-2)^2} = \sqrt{4x^2 + 8}$. $|\nabla f \cdot \mathbf{p}| = 2 \Rightarrow S =$

$$\iint_R \frac{|\nabla f|}{|\nabla f \cdot \mathbf{p}|}\,dA = \iint_R \frac{\sqrt{4x^2 + 8}}{2}\,dx\,dy = \int_0^2\int_0^{3x} \sqrt{x^2 + 2}\;dy\,dx = \int_0^2 3x\sqrt{x^2 + 2}\,dx = 6\sqrt{6} - 2\sqrt{2}$$

9. $\mathbf{p} = \mathbf{i}$, $\nabla f = \mathbf{i} + 2y\,\mathbf{j} + 2z\,\mathbf{k} \Rightarrow |\nabla f| = \sqrt{1^2 + (2y)^2 + (2z)^2} = \sqrt{1 + 4y^2 + 4z^2}$. $|\nabla f \cdot \mathbf{p}| = 1 \Rightarrow$

$$S = \iint_R \frac{|\nabla f|}{|\nabla f \cdot \mathbf{p}|}\,dA = \iint_R \sqrt{1 + 4y^2 + 4z^2}\;dy\,dz = \int_0^{2\pi}\int_1^{2} \sqrt{1 + 4r^2\cos^2\theta + 4r^2\sin^2\theta}\;r\,dr\,d\theta$$

$$= \int_0^{2\pi}\int_1^{2} \sqrt{1 + 4r^2}\;r\,dr\,d\theta = \frac{17\pi\sqrt{17} - 5\pi\sqrt{5}}{6}$$

13. On the faces in the coordinate planes, $g(x,y,z) = 0 \Rightarrow$ the integral over these faces is 0.

On the face, $x = a$, $f(x,y,z) = x = a$ and $g(x,y,z) = g(a,y,z) = ayz \Rightarrow \nabla f = \mathbf{i} \Rightarrow |\nabla f| = 1$. $\mathbf{p} = \mathbf{i} \Rightarrow |\nabla f \cdot \mathbf{p}| = 1$

$$\Rightarrow d\sigma = dy\,dz \Rightarrow \iint_{x=a} xyz\;d\sigma = \int_0^c\int_0^b ayz\;dy\,dz = \frac{ab^2c^2}{4}$$

On the face, $y = b$, $f(x,y,z) = y = b$ and $g(x,y,z) = g(x,b,z) = bxz \Rightarrow \nabla f = \mathbf{j} \Rightarrow |\nabla f| = 1$. $\mathbf{p} = \mathbf{j} \Rightarrow |\nabla f \cdot \mathbf{p}| = 1$

$$\Rightarrow d\sigma = dx\,dz \Rightarrow \iint_{y=b} xyz\;dx\,dz = \int_0^c\int_0^a bxz\;dz\,dx = \frac{a^2bc^2}{4}$$

On the face, $z = c$, $f(x,y,z) = z = c$ and $g(x,y,z) = g(x,y,c) = cxy \Rightarrow \nabla f = \mathbf{k} \Rightarrow |\nabla f| = 1$. $\mathbf{p} = \mathbf{k} \Rightarrow |\nabla f \cdot \mathbf{p}| = 1$

$$\Rightarrow d\sigma = dy\,dx \Rightarrow \iint_{z=c} xyz\;d\sigma = \int_0^b\int_0^a cxy\;dx\,dy = \frac{a^2b^2c}{4}$$

$$\therefore \iint_S g(x,y,z)\;d\sigma = \frac{abc(ab + ac + bc)}{4}$$

17. $\nabla G = 2x\,\mathbf{i} + 2y\,\mathbf{j} + 2z\,\mathbf{k} \Rightarrow |\nabla G| = \sqrt{4x^2 + 4y^2 + 4z^2} = 2a$. $\mathbf{n} = \dfrac{2x\,\mathbf{i} + 2y\,\mathbf{j} + 2z\,\mathbf{k}}{2\sqrt{x^2 + y^2 + z^2}} = \dfrac{x\,\mathbf{i} + y\,\mathbf{j} + z\,\mathbf{k}}{a} \Rightarrow$

$\mathbf{F} \cdot \mathbf{n} = \dfrac{z^2}{a}$. $|\nabla G \cdot \mathbf{k}| = 2z \Rightarrow d\sigma = \dfrac{2a}{2z}\,dA = \dfrac{a}{z}\,dA$. $\therefore$ Flux $= \displaystyle\int_R\int \frac{z^2}{a}\left(\frac{a}{z}\right) dA = \int_R\int z\, dA =$

$$\int_R\int \sqrt{a^2 - (x^2 + y^2)}\; dx\,dy = \int_0^{\pi/2}\int_0^a \sqrt{a^2 - r^2}\; r\,dr\,d\theta = \frac{a^3\pi}{6}$$

21. $\mathbf{n} = \dfrac{x\,\mathbf{i} + y\,\mathbf{j} + z\,\mathbf{k}}{a}$, $d\sigma = \dfrac{a}{z}\,dA$ (See Exercise 15) and $\mathbf{F} \cdot \mathbf{n} = \dfrac{x^2}{a} + \dfrac{y^2}{a} + \dfrac{z^2}{a} = a \Rightarrow$ Flux $= \displaystyle\int_R\int a\left(\frac{a}{z}\right) dA$

$$= \int_R\int \frac{a^2}{z}\,dA = \int_R\int \frac{a^2}{\sqrt{a^2 - (x^2 + y^2)}}\,dA = \int_0^{\pi/2}\int_0^a \frac{a^2}{\sqrt{a^2 - r^2}}\,r\,dr\,d\theta = \frac{a^3\pi}{2}$$

25. $\nabla G = -e^x\,\mathbf{i} + \mathbf{j} \Rightarrow |\nabla G| = \sqrt{e^{2x} + 1}$. $\mathbf{p} = \mathbf{i} \Rightarrow |\nabla G \cdot \mathbf{i}| = e^x$. $\mathbf{n} = \dfrac{e^x\,\mathbf{i} - \mathbf{j}}{\sqrt{e^{2x} + 1}} \Rightarrow \mathbf{F} \cdot \mathbf{n} = \dfrac{-2e^x - 2y}{\sqrt{e^{2x} + 1}}$.

$d\sigma = \dfrac{\sqrt{e^{2x} + 1}}{e^x}\,dA \Rightarrow$ Flux $= \displaystyle\int_R\int \frac{-2e^x - 2y}{\sqrt{e^{2x} + 1}}\left(\frac{\sqrt{e^{2x} + 1}}{e^x}\right) dA = \int_R\int -4\,dA = \int_0^1\int_1^2 -4\,dy\,dz$

$= -4$

29. $\nabla F = 2x\,\mathbf{i} + 2y\,\mathbf{j} + 2z\,\mathbf{k} \Rightarrow |\nabla F| = \sqrt{4x^2 + 4y^2 + 4z^2} = 2a$, $a > 0$ $\mathbf{p} = \mathbf{k} \Rightarrow |\nabla F \cdot \mathbf{k}| = 2z$ since $z \geq 0 \Rightarrow d\sigma =$

$\dfrac{2a}{2z}\,dA = \dfrac{a}{z}\,dA$. $\therefore$ $M = \displaystyle\int_S\int \delta\,d\sigma = \frac{\pi a^2}{2}\delta$. $M_{xy} = \displaystyle\int_S\int z\delta\,d\sigma = \delta\int_S\int z\left(\frac{a}{z}\right) dA =$

$a\delta\displaystyle\int_0^a\int_0^{\sqrt{a^2 - x^2}} dy\,dx = \frac{\pi a^3}{4}\delta$. $\therefore$ $\bar{z} = \dfrac{\frac{\pi a^3}{4}\delta}{\frac{\pi a^2}{2}\delta} = \dfrac{a}{2}$. Because of symmetry, $\bar{x} = \bar{y} = \dfrac{a}{2}$. $\therefore$ Centroid $= \left(\dfrac{a}{2}, \dfrac{a}{2}, \dfrac{a}{2}\right)$

33. $f_x(x,y) = 2x$, $f_y(x,y) = 2y \Rightarrow \sqrt{f_x^2 + f_y^2 + 1} = \sqrt{4x^2 + 4y^2 + 1} \Rightarrow$

Area $= \displaystyle\int_R\int \sqrt{4x^2 + 4y^2 + 1}\;dx\,dy = \int_0^{2\pi}\int_0^{\sqrt{3}} \sqrt{4r^2 + 1}\;r\,dr\,d\theta = \frac{\pi}{6}\left(13\sqrt{13} - 1\right)$

37. $y = \dfrac{2}{3}z^{3/2} \Rightarrow f_x(x,z) = 0$, $f_z(x,z) = z^{1/2} \Rightarrow \sqrt{f_x^2 + f_z^2 + 1} = \sqrt{z + 1} \Rightarrow$ Area $= \displaystyle\int_0^4\int_0^1 \sqrt{z + 1}\;dx\,dz =$

$$\int_0^4 \sqrt{z + 1}\,dz = \frac{2}{3}\left(5\sqrt{5} - 1\right)$$

15.6 PARAMETRIZED SURFACES

1. In cylindrical coordinates, let $x = r\cos\theta$, $y = r\sin\theta$, $z = \left(\sqrt{x^2+y^2}\right)^2 = r^2$. Then $\mathbf{r}(r,\theta) = (r\cos\theta)\,\mathbf{i} + (r\sin\theta)\,\mathbf{j} + r^2\,\mathbf{k}$, $0 \le r \le 2$, $0 \le \theta \le 2\pi$.

5. In cylindrical coordinates, let $x = r\cos\theta$, $y = r\sin\theta$, $z = \sqrt{x^2+y^2} = r$. Then $\mathbf{r}(r,\theta) = (r\cos\theta)\,\mathbf{i} + (r\sin\theta)\,\mathbf{j} + \sqrt{9-r^2}\,\mathbf{k}$ since $x^2+y^2+z^2 = 9 \Rightarrow z^2 = 9-(x^2+y^2) = 9-r^2 \Rightarrow z = \sqrt{9-r^2}$ $\left(z \ge 0 \text{ since } z = \sqrt{x^2+y^2}\right)$. Let $0 \le \theta \le 2\pi$. For the domain of r: $z = \sqrt{x^2+y^2}$ and $x^2+y^2+z^2 = 9 \Rightarrow x^2+y^2+\left(\sqrt{x^2+y^2}\right)^2 = 9 \Rightarrow 2\left(x^2+y^2\right) = 9 \Rightarrow 2r^2 = 9 \Rightarrow r = \sqrt{9/2}$ $(r \ge 0) = \frac{3}{\sqrt{2}}$. So, $0 \le r \le \frac{3}{\sqrt{2}}$.

9. Since $z = 4-y^2$, we can let $\mathbf{r}$ be a function of x and y $\Rightarrow \mathbf{r}(x,y) = x\,\mathbf{i} + y\,\mathbf{j} + (4-y^2)\,\mathbf{k}$. Let $0 \le x \le 2$. $z = 0 \Rightarrow 0 = 4-y^2 \Rightarrow y = \pm 2$. $\therefore$ let $-2 \le y \le 2$.

13. a) $x+y+z = 1 \Rightarrow z = 1-x-y$. In cylindrical coordinates, let $x = r\cos\theta$, $y = r\sin\theta$, $z = \sqrt{x^2+y^2}$. Then $\mathbf{r}(r,\theta) = r\cos\theta\,\mathbf{i} + r\sin\theta\,\mathbf{j} + (1 - r\cos\theta - r\sin\theta)\,\mathbf{k}$, $0 \le \theta \le 2\pi$, $0 \le r \le 3$.

b) In a fashion similar to cylindrical coordinates, but working in the yz-plane instead of the xy-plane, let $y = u\cos v$, $z = u\sin v$ where $u = \sqrt{y^2+z^2}$ and v is the angle formed by (x,y,z), (x,0,0), and (x,y,0) with (x,0,0) as vertex. Then $\mathbf{r}$ is a function of u and v $\Rightarrow \mathbf{r}(u,v) = (1 - u\cos v - u\sin v)\,\mathbf{i} + u\cos v\,\mathbf{j} + u\sin v\,\mathbf{k}$ (since $x+y+z = 1 \Rightarrow x = 1-y-z$), $0 \le u \le 3$, $0 \le v \le 2\pi$.

17. Let $x = r\cos\theta$, $y = r\sin\theta$, $z = r^2$. Then $\mathbf{r}(r,\theta) = r\cos\theta\,\mathbf{i} + r\sin\theta\,\mathbf{j} + \left(\frac{2 - r\sin\theta}{2}\right)\mathbf{k}$, $0 \le r \le 1$, $0 \le \theta \le 2\pi$.

$\mathbf{r}_r = \cos\theta\,\mathbf{i} + \sin\theta\,\mathbf{j} - \frac{\sin\theta}{2}\mathbf{k}$. $\mathbf{r}_\theta = -r\sin\theta\,\mathbf{i} + r\cos\theta\,\mathbf{j} - \frac{r\cos\theta}{2}\mathbf{k} \Rightarrow$

$$\mathbf{r}_r \times \mathbf{r}_\theta = \begin{vmatrix} \mathbf{i} & \mathbf{j} & \mathbf{k} \\ \cos\theta & \sin\theta & -\frac{\sin\theta}{2} \\ -r\sin\theta & r\cos\theta & -\frac{r\cos\theta}{2} \end{vmatrix} = \left(\frac{-r\sin\theta\cos\theta}{2} - \frac{-\sin\theta\,(r\cos\theta)}{2}\right)\mathbf{i} +$$

$$\left(\frac{r\sin^2\theta}{2} - \left(\frac{-r\cos^2\theta}{2}\right)\right)\mathbf{j} + \left(r\cos^2\theta - (-r\sin^2\theta)\right)\mathbf{k} = \frac{r}{2}\mathbf{j} + r\,\mathbf{k} \Rightarrow |\mathbf{r}_r \times \mathbf{r}_\theta| = \sqrt{\frac{r^2}{4} + r^2} = \frac{\sqrt{5}\,r}{2}.$$

$$\therefore A = \int_0^{2\pi}\!\!\int_0^1 \frac{\sqrt{5}\,r}{2}\,dr\,d\theta = \int_0^{2\pi} \left[\frac{\sqrt{5}\,r^2}{4}\right]_0^1 d\theta = \int_0^{2\pi} \frac{\sqrt{5}}{4}\,d\theta = \left[\frac{\sqrt{5}}{4}\theta\right]_0^{2\pi} = \frac{\pi\sqrt{5}}{2}.$$

21. Let $x = r\cos\theta$, $y = r\sin\theta \Rightarrow x^2+y^2 = r^2 = 1 \Rightarrow r = 1$. Then $\mathbf{r}(z,\theta) = \cos\theta\,\mathbf{i} + \sin\theta\,\mathbf{j} + 2\mathbf{k}$

$1 \le z \le 4$, $0 \le \theta \le 2\pi$. $\mathbf{r}_z = \mathbf{k}$ and $\mathbf{r}_\theta = -\sin\theta\,\mathbf{i} + \cos\theta\,\mathbf{j} \Rightarrow \mathbf{r}_\theta \times \mathbf{r}_z = \begin{vmatrix} \mathbf{i} & \mathbf{j} & \mathbf{k} \\ -\sin\theta & \cos\theta & 0 \\ 0 & 0 & 1 \end{vmatrix}$

$$= \cos\theta\,\mathbf{i} + \sin\theta\,\mathbf{j} \Rightarrow |\mathbf{r}_\theta \times \mathbf{r}_z| = \sqrt{\cos^2\theta + \sin^2\theta} = 1. \quad \therefore A = \int_0^{2\pi}\!\!\int_1^4 1\,dr\,d\theta = \int_0^{2\pi} 3\,d\theta = 6\pi.$$

25. Let $x = \rho \sin\phi \cos\theta$, $y = \rho \sin\phi \sin\theta$, $z = \rho\cos\phi$, $\rho = \sqrt{x^2 + y^2 + z^2} \Rightarrow \rho = \sqrt{2}$ on the sphere. $x^2 + y^2 + z^2 = 2$ and $z = \sqrt{x^2 + y^2} \Rightarrow z^2 + z^2 = 2 \Rightarrow z^2 = 1 \Rightarrow z = 1 \left(z = \sqrt{x^2 + y^2} \Rightarrow z \ge 0\right)$. $z = 1 \Rightarrow \phi = \frac{\pi}{4}$. For the lower portion of the sphere cut by the cone, $z = -\sqrt{2}$ when $x = 0$ and $y = 0 \Rightarrow \phi = \pi$. Then $\mathbf{r}(\phi,\theta) = \sqrt{2}\sin\phi\cos\theta\,\mathbf{i} + \sqrt{2}\sin\phi\sin\theta\,\mathbf{j} + \sqrt{2}\cos\phi\,\mathbf{k}$, $\frac{\pi}{4} \le \phi \le \pi$, $0 \le \theta \le 2\pi$. $\mathbf{r}_\phi = \sqrt{2}\cos\phi\cos\theta\,\mathbf{i} + \sqrt{2}\cos\phi\sin\theta\,\mathbf{j} - \sqrt{2}\sin\phi\,\mathbf{k}$ and $\mathbf{r}_\theta = -\sqrt{2}\sin\phi\sin\theta\,\mathbf{i} + \sqrt{2}\sin\phi\cos\theta\,\mathbf{j} \Rightarrow$

$$\mathbf{r}_\phi \times \mathbf{r}_\theta = \begin{vmatrix} \mathbf{i} & \mathbf{j} & \mathbf{k} \\ \sqrt{2}\cos\phi\cos\theta & \sqrt{2}\cos\phi\sin\theta & -\sqrt{2}\sin\phi \\ -\sqrt{2}\sin\phi\sin\theta & \sqrt{2}\sin\phi\cos\theta & 0 \end{vmatrix} = 2\sin^2\phi\cos\theta\,\mathbf{i} + 2\sin^2\phi\sin\theta\,\mathbf{j} +$$

$2\sin\phi\cos\phi\,\mathbf{k} \Rightarrow |\mathbf{r}_\phi \times \mathbf{r}_\theta| = \sqrt{4\sin^4\phi\cos^2\theta + 4\sin^4\phi\sin^2\theta + 4\sin^2\phi\cos^2\phi} = \sqrt{4\sin^2\phi} = 2\sin\phi$.

$$\therefore A = \int_0^{2\pi}\int_{\pi/4}^{\pi} 2\sin\phi\, d\phi\, d\theta = \int_0^{2\pi}\left(2 + \sqrt{2}\right) d\theta = \left(4 + 2\sqrt{2}\right)\pi.$$

29. Let the parametrization be $\mathbf{r}(\phi,\theta) = \sin\phi\cos\theta\,\mathbf{i} + \sin\phi\sin\theta\,\mathbf{j} + \cos\phi\,\mathbf{k}$ (spherical coordinates with $\rho = 1$ on the sphere), $0 \le \phi \le \pi$, $0 \le \theta \le 2\pi$. $\mathbf{r}_\phi = \cos\phi\cos\theta\,\mathbf{i} + \cos\phi\sin\theta\,\mathbf{j} - \sin\phi\,\mathbf{k}$ and $\mathbf{r}_\theta = -\sin\phi\sin\theta\,\mathbf{i} + \sin\phi\cos\theta\,\mathbf{j} \Rightarrow$

$$\mathbf{r}_\phi \times \mathbf{r}_\theta = \begin{vmatrix} \mathbf{i} & \mathbf{j} & \mathbf{k} \\ \cos\phi\cos\theta & \cos\phi\sin\theta & -\sin\phi \\ -\sin\phi\sin\theta & \sin\phi\cos\theta & 0 \end{vmatrix} = \sin^2\phi\cos\theta\,\mathbf{i} + \sin^2\phi\sin\theta\,\mathbf{j} + \sin\phi\cos\phi\,\mathbf{k} \Rightarrow |\mathbf{r}_\phi \times \mathbf{r}_\theta|$$

$= \sqrt{\sin^4\phi\cos^2\theta + \sin^4\phi\sin^2\theta + \sin^2\phi\cos^2\phi} = \sin\phi$. $\therefore$ since $x = \sin\phi\cos\theta \Rightarrow G(x,y,z) = G(\phi,\theta) = \cos^2\theta\sin^2\phi$, $\iint_S G(x,y,z)\, d\sigma = \int_0^{2\pi}\int_0^{\pi} (\cos^2\theta\sin^2\phi)(\sin\phi)\, d\phi\, d\theta = \frac{4\pi}{3}$. (Hint: Write $\sin^2\phi$ as $1 - \cos^2\phi$ for the first integration.)

33. Let $x = r\cos\theta$, $y = r\sin\theta$, $z = 1 - \left(\sqrt{x^2 + y^2}\right)^2 = 1 - r^2$ (cylindrical coordinates), $0 \le r \le 1$ since $0 \le z \le 1$, $0 \le \theta \le 2\pi \Rightarrow$ $\mathbf{r}(r,\theta) = r\cos\theta\,\mathbf{i} + r\sin\theta\,\mathbf{j} + (1 - r^2)\,\mathbf{k} \Rightarrow \mathbf{r}_r = \cos\theta\,\mathbf{i} + \sin\theta\,\mathbf{j} - 2r\,\mathbf{k}$ and $\mathbf{r}_\theta = -r\sin\theta\,\mathbf{i} + r\cos\theta\,\mathbf{j} \Rightarrow \mathbf{r}_r \times \mathbf{r}_\theta =$

$$\begin{vmatrix} \mathbf{i} & \mathbf{j} & \mathbf{k} \\ \cos\theta & \sin\theta & -2r \\ -r\sin\theta & r\cos\theta & 0 \end{vmatrix} = 2r^2\cos\theta\,\mathbf{i} + 2r^2\sin\theta\,\mathbf{j} + r\,\mathbf{k} \Rightarrow |\mathbf{r}_r \times \mathbf{r}_\theta| = \sqrt{(2r^2\cos\theta)^2 + (2r^2\sin\theta)^2 + (r)^2}$$

$= r\sqrt{1 + 4r^2}$. Since $z = 1 - r^2$ and $x = r\cos\theta$, $H(x,y,z) = H(r,\theta) = r^2\cos^2\theta\sqrt{1 + 4r^2} \Rightarrow \iint_S H(x,y,z)\, d\sigma =$

$$\int_0^{2\pi}\int_0^{1} \left(r^2\cos^2\theta\sqrt{1 + 4r^2}\right)\left(r\sqrt{1 + 4r^2}\right) dr\, d\theta = \int_0^{2\pi}\int_0^{1} \cos^2\theta\, r^3(1 + 4r^2)\, dr\, d\theta = \frac{11\pi}{12}$$

37. Let the parametrization be $\mathbf{r}(\phi,\theta) = a\sin\phi\cos\theta\,\mathbf{i} + a\sin\phi\sin\theta\,\mathbf{j} + a\cos\phi\,\mathbf{k}$ (spherical coordinates with $\rho = a, a \ge 0$, on the sphere), $0 \le \phi \le \frac{\pi}{2}, 0 \le \theta \le \frac{\pi}{2}$ (for the first octant). $\mathbf{r}_\phi = a\cos\phi\cos\theta\,\mathbf{i} + a\cos\phi\sin\theta\,\mathbf{j} - a\sin\phi\,\mathbf{k}$ and

$\mathbf{r}_\theta = -a\sin\phi\sin\theta\,\mathbf{i} + a\sin\phi\cos\theta\,\mathbf{j} \Rightarrow \mathbf{r}_\phi \times \mathbf{r}_\theta = \begin{vmatrix} \mathbf{i} & \mathbf{j} & \mathbf{k} \\ a\cos\phi\cos\theta & a\cos\phi\sin\theta & -a\sin\phi \\ -a\sin\phi\sin\theta & a\sin\phi\cos\theta & 0 \end{vmatrix} =$

$a^2\sin^2\phi\cos\theta\,\mathbf{i} + a^2\sin^2\phi\sin\theta\,\mathbf{j} + a^2\sin\phi\cos\phi\,\mathbf{k} \Rightarrow \mathbf{F}\cdot\mathbf{n}\,d\sigma = \mathbf{F}\cdot\frac{\mathbf{r}_\phi \times \mathbf{r}_\theta}{|\mathbf{r}_\phi \times \mathbf{r}_\theta|}|\mathbf{r}_\phi \times \mathbf{r}_\theta|\,d\theta\,d\phi = a^3\sin\phi\cos^2\phi$

since $\mathbf{F} = z\,\mathbf{k} = a\cos\phi\,\mathbf{k} \Rightarrow \int\int_S \mathbf{F}\cdot\mathbf{n}\,d\sigma = \int_0^{\pi/2}\int_0^{\pi/2} a^3\sin\phi\cos^2\phi\,d\phi\,d\theta = \frac{a^3\pi}{6}$

41. Let $x = r\cos\theta$, $y = r\sin\theta$, $z = \sqrt{x^2+y^2} = r$ (cylindrical coordinates), $0 \le r \le 1$ since $0 \le z \le 1$, $0 \le \theta \le 2\pi \Rightarrow$ $\mathbf{r}(r,\theta) = r\cos\theta\,\mathbf{i} + r\sin\theta\,\mathbf{j} + r\,\mathbf{k} \Rightarrow \mathbf{r}_r = \cos\theta\,\mathbf{i} + \sin\theta\,\mathbf{j} + \mathbf{k}$ and $\mathbf{r}_\theta = -r\sin\theta\,\mathbf{i} + r\cos\theta\,\mathbf{j} \Rightarrow \mathbf{r}_\theta \times \mathbf{r}_r =$

$\begin{vmatrix} \mathbf{i} & \mathbf{j} & \mathbf{k} \\ -r\sin\theta & r\cos\theta & 0 \\ \cos\theta & \sin\theta & 1 \end{vmatrix} = r\cos\theta\,\mathbf{i} + r\sin\theta\,\mathbf{j} - r\,\mathbf{k} \Rightarrow \mathbf{F}\cdot\mathbf{n}\,d\sigma\,\mathbf{F}\cdot\frac{\mathbf{r}_\theta \times \mathbf{r}_r}{|\mathbf{r}_\theta \times \mathbf{r}_r|}|\mathbf{r}_\theta \times \mathbf{r}_r|\,d\theta\,dr =$

$\left(r^3\sin\theta\cos^2\theta + r^2\right)d\theta\,dr$ since $\mathbf{F} = r^2\sin\theta\cos\theta\,\mathbf{i} - r\,\mathbf{k} \Rightarrow \int\int_S \mathbf{F}\cdot\mathbf{n}\,d\sigma = \int_0^{2\pi}\int_0^1 (r^3\sin\theta\cos^2\theta + r^2)\,dr\,d\theta$

$= \frac{2\pi}{3}$

45. Let the parametrization be $\mathbf{r}(\phi,\theta) = a\sin\phi\cos\theta\,\mathbf{i} + a\sin\phi\sin\theta\,\mathbf{j} + a\cos\phi\,\mathbf{k}$, $0 \le \phi \le \frac{\pi}{2}, 0 \le \theta \le \frac{\pi}{2}, \Rightarrow \mathbf{r}_\phi =$ $a\cos\phi\cos\theta\,\mathbf{i} + a\cos\phi\sin\theta\,\mathbf{j} - a\sin\phi\,\mathbf{k}$ and $\mathbf{r}_\theta = -a\sin\phi\sin\theta\,\mathbf{i} + a\sin\phi\cos\theta\,\mathbf{j} \Rightarrow \mathbf{r}_\phi \times \mathbf{r}_\theta =$

$\begin{vmatrix} \mathbf{i} & \mathbf{j} & \mathbf{k} \\ a\cos\phi\cos\theta & a\cos\phi\sin\theta & -a\sin\phi \\ -a\sin\phi\sin\theta & a\sin\phi\cos\theta & 0 \end{vmatrix} = a^2\sin^2\phi\cos\theta\,\mathbf{i} + a^2\sin^2\phi\sin\theta\,\mathbf{j} + a^2\sin\phi\cos\phi\,\mathbf{k} \Rightarrow$

$|\mathbf{r}_\phi \times \mathbf{r}_\theta| = \sqrt{a^4\sin^4\phi\cos^2\theta + a^4\sin^4\phi\sin^2\theta + a^4\sin^2\phi\cos^2} = \sqrt{a^4\sin^2\phi} = a^2\sin\phi$. The mass

$M = \int\int_S d\sigma = \int_0^{\pi/2}\int_0^{\pi/2} a^2\sin\phi\,d\phi\,d\theta = \frac{a^2\pi}{2}$. The first moment $M_{yz} = \int\int_S x\,d\sigma =$

$\int_0^{\pi/2}\int_0^{\pi/2} (a\sin\phi\cos\theta)(a^2\sin\phi)\,d\phi\,d\theta = \frac{a^3\pi}{4} \Rightarrow \bar{x} = \frac{a^3\pi/4}{a^2\pi/2} = \frac{a}{2}$. The centroid is located at $\left(\frac{a}{2}, \frac{a}{2}, \frac{a}{2}\right)$

by symmetry.

49.

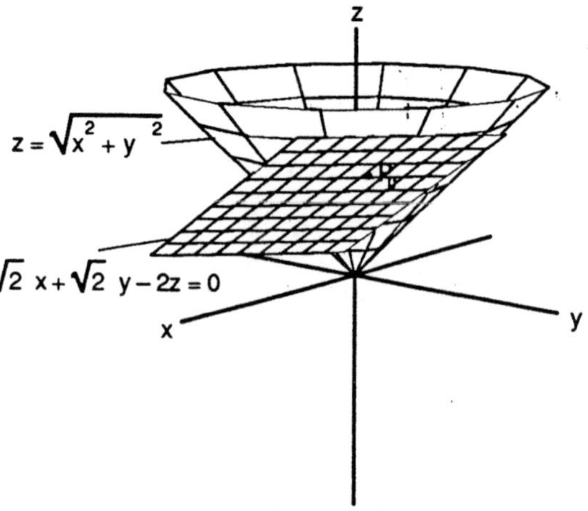

Graph 15.6.49

The parametrization $\mathbf{r}(r,\theta)$ at $P_0 = \left(\sqrt{2}, \sqrt{2}, 2\right)$ $\Rightarrow \theta = \frac{\pi}{4}, r = 2 \Rightarrow \mathbf{r}_r = \cos\theta\,\mathbf{i} + \sin\theta\,\mathbf{j} + \mathbf{k} = \frac{\sqrt{2}}{2}\mathbf{i} + \frac{\sqrt{2}}{2}\mathbf{j} + \mathbf{k}$ and $\mathbf{r}_\theta = -r\sin\theta\,\mathbf{i} + r\cos\theta\,\mathbf{j} = -\sqrt{2}\,\mathbf{i} + \sqrt{2}\,\mathbf{j} \Rightarrow \mathbf{r}_r \times \mathbf{r}_\theta = \begin{vmatrix} \mathbf{i} & \mathbf{j} & \mathbf{k} \\ \sqrt{2}/2 & \sqrt{2}/2 & 1 \\ -\sqrt{2} & \sqrt{2} & 0 \end{vmatrix} =$ $-\sqrt{2}\,\mathbf{i} - \sqrt{2}\,\mathbf{j} + 2\mathbf{k} \Rightarrow$ the tangent plane is $\left(-\sqrt{2}\,\mathbf{i} - \sqrt{2}\,\mathbf{j} + 2\mathbf{k}\right) \cdot \left(\left(x - \sqrt{2}\right)\mathbf{i} + \left(y - \sqrt{2}\right)\mathbf{j} + (z-2)\mathbf{k}\right) = \sqrt{2}\,x + \sqrt{2}\,y - 2z = 0.$ The parametrization $\mathbf{r}(r,\theta) \Rightarrow x = r\cos\theta,\ y = r\sin\theta$ and $z = r \Rightarrow x^2 + y^2 = r^2 = z^2 \Rightarrow$ the surface $z = \sqrt{x^2 + y^2}$

15.7 STOKE'S THEOREM

1. $\text{curl}\,\mathbf{F} = \nabla \times \mathbf{F} = 2\,\mathbf{k},\ \mathbf{n} = \mathbf{k} \Rightarrow \text{curl}\,\mathbf{F} \cdot \mathbf{n} = 2 \Rightarrow d\sigma = dx\,dy \Rightarrow \oint_C \mathbf{F} \cdot d\mathbf{r} = \iint_R 2\,dA =$ 2(Area of the ellipse) $= 4\pi$

5. $\text{curl}\,\mathbf{F} = \nabla \times \mathbf{F} = (2y - 0)\,\mathbf{i} + (2z - 2x)\,\mathbf{j} + (2x - 2y)\,\mathbf{k} = 2y\,\mathbf{i} + (2z - 2x)\,\mathbf{j} + (2x - 2y)\,\mathbf{k},$ $\mathbf{n} = \mathbf{k} \Rightarrow \text{curl}\,\mathbf{F} \cdot \mathbf{n} = 2x - 2y \Rightarrow d\sigma = dx\,dy \Rightarrow \oint_C \mathbf{F} \cdot d\mathbf{r} = \int_{-1}^{1}\int_{-1}^{1} (2x - 2y)\,dx\,dy = 0$

9. $\text{curl}\,\mathbf{F} = \nabla \times \mathbf{F} = -2x\,\mathbf{j} + 2\,\mathbf{k}$. Flux of $\nabla \times \mathbf{F} = \iint_S \nabla \times \mathbf{F} \cdot \mathbf{n}\,d\sigma = \oint_C \mathbf{F} \cdot d\mathbf{r}$. Let C be $x = a\cos t$, $y = a\sin t$. Then $\mathbf{r} = (a\cos t)\,\mathbf{i} + (a\sin t)\,\mathbf{j} \Rightarrow d\mathbf{r} = (-a\sin t)\,dt\,\mathbf{i} + (a\cos t)\,dt\,\mathbf{j}$. Then $\mathbf{F} \cdot d\mathbf{r} = ay\sin t\,dt + ax\cos t\,dt = a^2\sin^2 t\,dt + a^2\cos^2 t\,dt = a^2 dt$. $\therefore$ Flux of $\nabla \times \mathbf{F} = \oint_C \mathbf{F} \cdot d\mathbf{r} = \int_0^{2\pi} a^2\,dt = 2\pi a^2$

13. a) $\mathbf{F} = 2x\,\mathbf{i} + 2y\,\mathbf{j} + 2z\,\mathbf{k} \Rightarrow \text{curl } \mathbf{F} = \mathbf{0}$ $\therefore \oint_C \mathbf{F}\cdot d\mathbf{r} = \iint_S \nabla \times \mathbf{F}\cdot\mathbf{n}\, d\sigma = \iint_S d\sigma = 0$

b) $\mathbf{F} = \nabla(x^2y^2z^3)$. Let $f(x,y,z) = x^2y^2z^3 \Rightarrow \nabla\mathbf{F} = \nabla \times \nabla f = \mathbf{0} \Rightarrow \text{curl } \mathbf{F} = \mathbf{0}$

$\therefore \oint_C \mathbf{F}\cdot d\mathbf{r} = \iint_S \nabla \times \mathbf{F}\cdot\mathbf{n}\, d\sigma = \iint_S 0\, d\sigma = 0$

c) $\mathbf{F} = \nabla \times (x\,\mathbf{i} + y\,\mathbf{j} + z\,\mathbf{k}) = \mathbf{0} \Rightarrow \nabla \times \mathbf{F} = \mathbf{0}$. $\therefore \oint_C \mathbf{F}\cdot d\mathbf{r} = \iint_S \nabla \times \mathbf{F}\cdot\mathbf{n}\, d\sigma = \iint_S 0\, d\sigma = 0$

d) $\mathbf{F} = \nabla f \Rightarrow \nabla \times \mathbf{F} = \nabla \times \nabla f = \mathbf{0}$. $\therefore \oint_C \mathbf{F}\cdot d\mathbf{r} = \iint_S \nabla \times \mathbf{F}\cdot\mathbf{n}\, d\sigma = \iint_S 0\, d\sigma = 0$

17. Yes: If $\nabla \times \mathbf{F} = \mathbf{0}$, then the circulation of $\mathbf{F}$ around the boundary C of any oriented surface S in the domain of $\mathbf{F}$ is zero. The reason is: By Stoke's theorem, circulation $= \oint_C \mathbf{F}\cdot d\mathbf{r} = \iint_S \nabla \times \mathbf{F}\cdot\mathbf{n}\, d\sigma = \iint_S \mathbf{0}\cdot\mathbf{n}\, d\sigma = 0$

15.8 THE DIVERGENCE THEOREM

1. $\mathbf{F} = \dfrac{-y\,\mathbf{i} + x\,\mathbf{j}}{\sqrt{x^2 + y^2}} \Rightarrow \text{div } \mathbf{F} = \dfrac{xy - xy}{(x^2 + y^2)^{3/2}} = 0$

5. $\dfrac{\partial}{\partial x}(y - x) = -1, \dfrac{\partial}{\partial y}(z - y) = -1, \dfrac{\partial}{\partial z}(y - x) = 0 \Rightarrow \nabla\cdot\mathbf{F} = -2 \Rightarrow \text{Flux} = \int_{-1}^{1}\int_{-1}^{1}\int_{-1}^{1} -2\, dx\, dy\, dz = -16$

9. $\dfrac{\partial}{\partial x}(x^2) = 2x, \dfrac{\partial}{\partial y}(-2xy) = -2x, \dfrac{\partial}{\partial z}(3xz) = 3x \Rightarrow \text{Flux} = \iiint_D 3x\, dx\, dy\, dz =$

$$\int_0^{\pi/2}\int_0^{\pi/2}\int_0^{2} 3\rho \sin\phi \cos\theta(\rho^2 \sin\phi)\, d\rho\, d\phi\, d\theta = 3\pi$$

13. Let $\rho = \sqrt{x^2 + y^2 + z^2}$. Then $\dfrac{\partial\rho}{\partial x} = \dfrac{x}{\rho}, \dfrac{\partial\rho}{\partial y} = \dfrac{y}{\rho}, \dfrac{\partial\rho}{\partial z} = \dfrac{z}{\rho} \Rightarrow \dfrac{\partial}{\partial x}(\rho x) = \dfrac{\partial\rho}{\partial x}x + \rho = \dfrac{x^2}{\rho} = \rho, \dfrac{\partial}{\partial y}(\rho y) =$
$\dfrac{\partial\rho}{\partial y}y + \rho = \dfrac{y^2}{\rho} + \rho, \dfrac{\partial}{\partial z}(\rho z) = \dfrac{\partial\rho}{\partial z}z + \rho = \dfrac{z^2}{\rho} + \rho \Rightarrow \nabla\cdot\mathbf{F} = \dfrac{x^2 + y^2 + z^2}{\rho} + 3\rho = 4\rho$ since $\rho = \sqrt{x^2 + y^2 + z^2}$

$\Rightarrow \text{Flux} = \iiint_D 4\rho\, dV = \iiint_D 4\sqrt{x^2 + y^2 + z^2}\, dx\, dy\, dz =$

$$\int_0^{2\pi}\int_0^{\pi}\int_1^{\sqrt{2}} (4\rho)\rho^2 \sin\phi\, d\rho\, d\phi\, d\theta = 12\pi$$

17. $|\mathbf{F}\cdot\mathbf{n}| \le \|\mathbf{F}\|\,\|\mathbf{n}\| \le 1$ since $\|\mathbf{F}\| \le 1$, $\|\mathbf{n}\| = 1$. Then $\iiint_D \nabla\cdot\mathbf{F}\,d\sigma = \iint_S \mathbf{F}\cdot\mathbf{n}\,d\sigma \le$

$\iint_S |\mathbf{F}\cdot\mathbf{n}|\,d\sigma \le \iint_S 1\,d\sigma =$ Area of S

21. a) $\mathbf{G} = M\mathbf{i} + N\mathbf{j} + P\mathbf{k} \Rightarrow \text{curl}\,\mathbf{G} = \left(\frac{\partial P}{\partial y} - \frac{\partial N}{\partial z}\right)\mathbf{i} + \left(\frac{\partial M}{\partial z} - \frac{\partial P}{\partial x}\right)\mathbf{j} + \left(\frac{\partial N}{\partial x} - \frac{\partial M}{\partial y}\right)\mathbf{k}$.

$\nabla\cdot\nabla\times\mathbf{G} = \text{div}(\text{curl}\,\mathbf{G}) = \frac{\partial}{\partial x}\left(\frac{\partial P}{\partial y} - \frac{\partial N}{\partial z}\right) + \frac{\partial}{\partial y}\left(\frac{\partial M}{\partial z} - \frac{\partial P}{\partial x}\right) + \frac{\partial}{\partial z}\left(\frac{\partial N}{\partial x} - \frac{\partial M}{\partial y}\right) = \frac{\partial^2 P}{\partial x\partial y} - \frac{\partial^2 N}{\partial x\partial z} +$

$\frac{\partial^2 M}{\partial y\partial z} - \frac{\partial^2 P}{\partial y\partial x} + \frac{\partial^2 N}{\partial z\partial x} - \frac{\partial^2 M}{\partial z\partial y} = 0$ if the partial derivatives are continuous.

b) $\iint_S \nabla\times\mathbf{G}\cdot\mathbf{n}\,d\sigma = \iiint_D \nabla\cdot\nabla\times\mathbf{G}\,dV = \iiint_D 0\,dV = 0$ if the

divergence theorem applies.

23. a) Let $\mathbf{F}_1 = M_1\mathbf{i} + N_1\mathbf{j} + P_1\mathbf{k}$, $\mathbf{F}_2 = M_2\mathbf{i} + N_2\mathbf{j} + P_2\mathbf{k} \Rightarrow a\mathbf{F}_1 + b\mathbf{F}_2 = (aM_1 + bM_2)\,\mathbf{i} + (aN_1) + bN_2)\,\mathbf{j}\,+$

$(aP_1 + bP_2)\,\mathbf{k}$. Then $\nabla\cdot(a\mathbf{F}_1 + b\mathbf{F}_2) = \left(a\frac{\partial M_1}{\partial x} + b\frac{\partial M_2}{\partial x}\right) + \left(a\frac{\partial N_1}{\partial y} + b\frac{\partial N_2}{\partial y}\right) + \left(a\frac{\partial P_1}{\partial z} + b\frac{\partial P_2}{\partial z}\right) =$

$a\left(\frac{\partial M_1}{\partial x} + \frac{\partial N_1}{\partial y} + \frac{\partial P_1}{\partial z}\right) + b\left(\frac{\partial M_2}{\partial x} + \frac{\partial N_2}{\partial y} + \frac{\partial P_2}{\partial z}\right) = a(\nabla\cdot\mathbf{F}_1) + b(\nabla\cdot\mathbf{F}_2)$

b) Define $\mathbf{F}_1$ and $\mathbf{F}_2$ as in part a. Then $\nabla\times(a\mathbf{F}_1 + b\mathbf{F}_2) = \left(a\frac{\partial P_1}{\partial y} + b\frac{\partial P_2}{\partial y} - \left(a\frac{\partial N_1}{\partial z} + b\frac{\partial N_2}{\partial z}\right)\right)\mathbf{i}\,+$

$\left(a\frac{\partial M_1}{\partial z} + b\frac{\partial M_2}{\partial z} - \left(a\frac{\partial P_1}{\partial x} + b\frac{\partial P_2}{\partial x}\right)\right)\mathbf{j} + \left(a\frac{\partial N_1}{\partial x} + b\frac{\partial N_2}{\partial x} - \left(a\frac{\partial M_1}{\partial y} + b\frac{\partial M_2}{\partial y}\right)\right)\mathbf{k} =$

$a\left[\left(\frac{\partial P_1}{\partial y} - \frac{\partial N_1}{\partial z}\right)\mathbf{i} + \left(\frac{\partial M_1}{\partial z} - \frac{\partial P_1}{\partial x}\right)\mathbf{j} + \left(\frac{\partial N_1}{\partial x} - \frac{\partial M_1}{\partial y}\right)\mathbf{k}\right] + b\left[\left(\frac{\partial P_2}{\partial y} - \frac{\partial N_2}{\partial z}\right)\mathbf{i} + \left(\frac{\partial M_2}{\partial z} - \frac{\partial P_2}{\partial x}\right)\mathbf{j} + \left(\frac{\partial N_2}{\partial x} - \frac{\partial N_2}{\partial y}\right)\mathbf{k}\right]$

$= a\nabla\times\mathbf{F}_1 + b\nabla\times\mathbf{F}_2$

c) $\text{div}(g\mathbf{F}) = \nabla\cdot g\mathbf{F} = \frac{\partial}{\partial x}(gM) + \frac{\partial}{\partial y}(gN) + \frac{\partial}{\partial z}(gP) = \left(g\frac{\partial M}{\partial x} + M\frac{\partial g}{\partial x}\right) + \left(g\frac{\partial N}{\partial y} + N\frac{\partial g}{\partial y}\right) + \left(g\frac{\partial P}{\partial z} + P\frac{\partial g}{\partial z}\right) =$

$\left(M\frac{\partial g}{\partial x} + N\frac{\partial g}{\partial y} + P\frac{\partial g}{\partial z}\right) + g\left(\frac{\partial M}{\partial x} + \frac{\partial N}{\partial y} + \frac{\partial P}{\partial z}\right) = \mathbf{F}\cdot\nabla g + g(\nabla\cdot\mathbf{F})$

d) $\nabla\times(g\mathbf{F}) = \left(\frac{\partial}{\partial y}(gP) - \frac{\partial}{\partial z}(gN)\right)\mathbf{i} + \left(\frac{\partial}{\partial z}(gM) - \frac{\partial}{\partial x}(gP)\right)\mathbf{j} + \left(\frac{\partial}{\partial x}(gN) - \frac{\partial}{\partial y}(gM)\right)\mathbf{k} =$

$\left(P\frac{\partial g}{\partial y} + g\frac{\partial P}{\partial y} - N\frac{\partial g}{\partial z} - g\frac{\partial N}{\partial z}\right)\mathbf{i} + \left(M\frac{\partial g}{\partial z} + g\frac{\partial M}{\partial z} - P\frac{\partial g}{\partial x} - g\frac{\partial P}{\partial x}\right)\mathbf{j} + \left(N\frac{\partial g}{\partial x} + g\frac{\partial N}{\partial x} - M\frac{\partial g}{\partial y} - g\frac{\partial M}{\partial y}\right)\mathbf{k}$

$= \left(P\frac{\partial g}{\partial y} - N\frac{\partial g}{\partial z}\right)\mathbf{i} + \left(g\frac{\partial P}{\partial y} - g\frac{\partial N}{\partial z}\right)\mathbf{i} + \left(M\frac{\partial g}{\partial z} - P\frac{\partial g}{\partial x}\right)\mathbf{j} + \left(g\frac{\partial M}{\partial z} - g\frac{\partial P}{\partial x}\right)\mathbf{j} + \left(N\frac{\partial g}{\partial x} - M\frac{\partial g}{\partial y}\right)\mathbf{k}\,+$

$\left(g\frac{\partial N}{\partial x} - g\frac{\partial M}{\partial y}\right)\mathbf{k} = g(\nabla\times\mathbf{F}) + \nabla g\times\mathbf{F}$

e) $(\mathbf{F}_1 \times \mathbf{F}_2) = \begin{vmatrix} \mathbf{i} & \mathbf{j} & \mathbf{k} \\ M_1 & N_1 & P_1 \\ M_2 & N_2 & P_2 \end{vmatrix} = (N_1P_2 - P_1N_2)\,\mathbf{i} - (M_1P_2 - P_1M_2)\,\mathbf{j} + (M_1N_2 - N_1M_2)\,\mathbf{k} \Rightarrow$

$\nabla \cdot (\mathbf{F}_1 \times \mathbf{F}_2) = \nabla \cdot \left((N_1P_2 - P_1N_2)\,\mathbf{i} - (M_1P_2 - P_1M_2)\,\mathbf{j} + (M_1N_2 - N_1M_2)\,\mathbf{k} \right) =$

$\frac{\partial}{\partial x}(N_1P_2 - P_1N_2) - \frac{\partial}{\partial y}(M_1P_2 - P_1M_2) + \frac{\partial}{\partial z}(M_1N_2 - N_1M_2) = P_2\frac{\partial N_1}{\partial x} + N_1\frac{\partial P_2}{\partial x} - N_2\frac{\partial P_1}{\partial x} - P_1\frac{\partial N_2}{\partial x}$

$- M_1\frac{\partial P_2}{\partial y} - P_2\frac{\partial M_1}{\partial y} + P_1\frac{\partial M_2}{\partial y} + M_2\frac{\partial P_1}{\partial y} + M_1\frac{\partial N_2}{\partial z} + N_2\frac{\partial M_1}{\partial z} - N_1\frac{\partial M_2}{\partial z} - M_2\frac{\partial N_1}{\partial z} =$

$M_2\left(\frac{\partial P_1}{\partial y} - \frac{\partial N_1}{\partial z}\right) + N_2\left(\frac{\partial M_1}{\partial z} - \frac{\partial P_1}{\partial x}\right) + P_2\left(\frac{\partial N_1}{\partial x} - \frac{\partial M_1}{\partial y}\right) + M_1\left(\frac{\partial N_2}{\partial z} - \frac{\partial P_2}{\partial y}\right) + N_1\left(\frac{\partial P_2}{\partial x} - \frac{\partial M_2}{\partial z}\right) +$

$P_1\left(\frac{\partial M_2}{\partial y} - \frac{\partial N_2}{\partial x}\right) = \mathbf{F}_2 \cdot (\nabla \times \mathbf{F}_1) - \mathbf{F}_1 \cdot (\nabla \times \mathbf{F}_2)$

15.P PRACTICE EXERCISES

1. Path 1: $\mathbf{r} = t\,\mathbf{i} + t\,\mathbf{j} + t\,\mathbf{k} \Rightarrow x = t,\ y = t,\ z = t,\ 0 \le t \le 1 \Rightarrow f(g(t),h(t),k(t)) = 3 - 3t^2$ and $\frac{dx}{dt} = 1, \frac{dy}{dt} = 1,$

$\frac{dz}{dt} = 1 \Rightarrow \sqrt{\left(\frac{dx}{dt}\right)^2 + \left(\frac{dy}{dt}\right)^2 + \left(\frac{dz}{dt}\right)^2}\,dt = \sqrt{3}\,dt \Rightarrow \int_C f(x,y,z)\,ds = \int_0^1 \sqrt{3}\left(3 - 3t^2\right)dt = 2\sqrt{3}$

Path 2: $\mathbf{r}_1 = t\,\mathbf{i} + t\,\mathbf{j},\ 0 \le t \le 1 \Rightarrow x = t,\ y = t,\ z = 0 \Rightarrow f(g(t),h(t),k(t)) = 2t - 3t^2 + 3$ and $\frac{dx}{dt} = 1, \frac{dy}{dt} = 1,$

$\frac{dz}{dt} = 0 \Rightarrow \sqrt{\left(\frac{dx}{dt}\right)^2 + \left(\frac{dy}{dt}\right)^2 + \left(\frac{dz}{dt}\right)^2}\,dt = \sqrt{2}\,dt \Rightarrow \int_{C_1} f(x,y,z)\,ds = \int_0^1 \sqrt{2}\left(2t - 3t^2 + 3\right)dt =$

$3\sqrt{2}$. $\mathbf{r}_2 = \mathbf{i} + \mathbf{j} + t\,\mathbf{k} \Rightarrow x = 1,\ y = 1,\ z = t \Rightarrow f(g(t),h(t),k(t)) = 2 - 2t$ and $\frac{dx}{dt} = 0, \frac{dy}{dt} = 0, \frac{dz}{dt} = 1 \Rightarrow$

$\sqrt{\left(\frac{dx}{dt}\right)^2 + \left(\frac{dy}{dt}\right)^2 + \left(\frac{dz}{dt}\right)^2}\,dt = dt \Rightarrow \int_{C_2} f(x,y,z)\,ds = \int_0^1 (2 - 2t)\,dt = 1.\ \therefore\ \int_C f(x,y,z)\,ds =$

$\int_{C_1} f(x,y,z)\,ds + \int_{C_2} f(x,y,z)\,ds = 3\sqrt{2} + 1$

5. a) $\mathbf{r} = \sqrt{2}\,t\,\mathbf{i} + \sqrt{2}\,t\,\mathbf{j} + (4 - t^2)\,\mathbf{k},\ 0 \le t \le 1 \Rightarrow x = \sqrt{2}\,t,\ y = \sqrt{2}\,t,\ z = 4 - t^2 \Rightarrow \frac{dx}{dt} = \sqrt{2}, \frac{dy}{dt} = \sqrt{2}, \frac{dz}{dt} = -2t$

$\Rightarrow \sqrt{\left(\frac{dx}{dt}\right)^2 + \left(\frac{dy}{dt}\right)^2 + \left(\frac{dz}{dt}\right)^2}\,dt = \sqrt{4 + 4t^2}\,dt \Rightarrow M = \int_C \delta(x,y,z)\,ds = \int_0^1 3t\sqrt{4 + 4t^2}\,dt =$

$4\sqrt{2} - 2$

b) $M = \int_C \delta(x,y,z)\,ds = \int_0^1 \sqrt{4 + 4t^2}\,dt = \sqrt{2} + \ln(1 + \sqrt{2})$

9. a) $x^2 + y^2 = 1 \Rightarrow \mathbf{r} = (\cos t)\,\mathbf{i} + (\sin t)\,\mathbf{j},\ 0 \le t \le \pi \Rightarrow x = \cos t,\ y = \sin t \Rightarrow \mathbf{F} = (\cos t + \sin t)\,\mathbf{i} - \mathbf{j}$ and $\frac{d\mathbf{r}}{dt} = (-\sin t)\,\mathbf{i} + (\cos t)\,\mathbf{j} \Rightarrow \mathbf{F} \cdot \frac{d\mathbf{r}}{dt} = -\sin t \cos t - \sin^2 t - \cos t \Rightarrow$ Flow $=$

$$\int_0^{\pi} (-\sin t \cos t - \sin^2 t - \cos t)\,dt = -\frac{1}{2}\pi$$

b) $\mathbf{r} = -t\,\mathbf{i},\ -1 \le t \le 1 \Rightarrow x = -t,\ y = 0 \Rightarrow \mathbf{F} = -t\,\mathbf{i} - t^2\,\mathbf{j}$ and $\frac{d\mathbf{r}}{dt} = -\mathbf{i} \Rightarrow \mathbf{F} \cdot \frac{d\mathbf{r}}{dt} = t \Rightarrow$ Flow $= \int_{-1}^{1} t\,dt = 0$

c) $\mathbf{r}_1 = (1-t)\,\mathbf{i} - t\,\mathbf{j},\ 0 \le t \le 1 \Rightarrow \mathbf{F}_1 = (1-2t)\,\mathbf{i} - (1 - 2t - 2t^2)\,\mathbf{j}$ and $\frac{d\mathbf{r}_1}{dt} = -\mathbf{i} - \mathbf{j} \Rightarrow \mathbf{F}_1 \cdot \frac{d\mathbf{r}_1}{dt} = 2t^2 \Rightarrow$ $\text{Flow}_1 = \int_0^1 2t^2\,dt = \frac{2}{3}$. $\mathbf{r}_2 = -t\,\mathbf{i} + (t-1)\,\mathbf{j},\ 0 \le t \le 1 \Rightarrow \mathbf{F}_2 = -\mathbf{i} - (2t^2 - 2t + 1)\,\mathbf{j}$ and $\frac{d\mathbf{r}_2}{dt} = -\mathbf{i} + \mathbf{j} \Rightarrow \mathbf{F}_2 \cdot \frac{d\mathbf{r}_2}{dt} = -2t^2 + 2t \Rightarrow \text{Flow}_2 = \int_0^1 (-2t^2 + 2t)\,dt = \frac{1}{3}$. $\therefore$ Flow $= \text{Flow}_1 + \text{Flow}_2 = 1$

13. $\frac{\partial P}{\partial y} = 0 \ne ye^z = \frac{\partial N}{\partial z} \Rightarrow$ Not Conservative

17. $\frac{\partial P}{\partial y} = -\frac{1}{2}(x+y+z)^{-3/2} = \frac{\partial N}{\partial z},\ \frac{\partial M}{\partial z} = -\frac{1}{2}(x+y+z)^{-3/2} = \frac{\partial P}{\partial x},\ \frac{\partial N}{\partial x} = -\frac{1}{2}(x+y+z)^{-3/2} = \frac{\partial M}{\partial y} \Rightarrow M\,dx + N\,dy + P\,dz$ is exact. $\frac{\partial f}{\partial x} = \frac{1}{\sqrt{x+y+z}} \Rightarrow f(x,y,z) = 2\sqrt{x+y+z} + g(y,z)$. $\frac{\partial f}{\partial y} = \frac{1}{\sqrt{x+y+z}} + \frac{\partial g}{\partial y} = \frac{1}{\sqrt{x+y+z}}$ $\Rightarrow \frac{\partial g}{\partial y} = 0 \Rightarrow g(y,z) = h(z) \Rightarrow f(x,y,z) = 2\sqrt{x+y+z} + h(z)$. $\frac{\partial f}{\partial z} = \frac{1}{\sqrt{x+y+z}} + h'(z) = \frac{1}{\sqrt{x+y+z}} \Rightarrow h'(z) = 0$ $\Rightarrow h(z) = C \Rightarrow f(x,y,z) = 2\sqrt{x+y+z} + C \Rightarrow$

$$\int_{(-1,1,1)}^{(4,-3,0)} \frac{dx + dy + dz}{\sqrt{x+y+z}} = f(4,-3,0) - f(-1,1,1) = 0$$

21. $\mathbf{F} = x\,\mathbf{i} + y\,\mathbf{j} + z\,\mathbf{k} \Rightarrow \frac{\partial P}{\partial y} = 0 = \frac{\partial N}{\partial z},\ \frac{\partial M}{\partial z} = 0 = \frac{\partial P}{\partial x},\ \frac{\partial N}{\partial x} = 0 = \frac{\partial M}{\partial y} \Rightarrow \mathbf{F}$ is conservative $\Rightarrow$ curl $\mathbf{F} = \mathbf{0} \Rightarrow$ circulation is 0.

25. $M = 2xy + x,\ N = xy - y \Rightarrow \frac{\partial M}{\partial x} = 2y + 1,\ \frac{\partial M}{\partial y} = 2x,\ \frac{\partial N}{\partial x} = y,\ \frac{\partial N}{\partial y} = x - 1 \Rightarrow$ Flux $=$

$$\iint_R (2y + 1 + x - 1)\,dy\,dx = \int_0^1\int_0^1 (2y + x)\,dy\,dx = \frac{3}{2}.\ \text{Circ} = \iint_R (y - 2x)\,dy\,dx =$$

$$\int_0^1\int_0^1 (y - 2x)\,dy\,dx = -\frac{1}{2}$$

29. Let $M = 8x \sin y$, $N = -8y \cos x \Rightarrow \frac{\partial M}{\partial y} = 8x \cos y$, $\frac{\partial N}{\partial x} = 8y \sin x \Rightarrow \int_C 8x \sin y \, dx - 8y \cos x \, dy =$

$$\iint_R (8y \sin x - 8x \cos y) \, dy \, dx = \int_0^{\pi/2} \int_0^{\pi/2} (8y \sin x - 8x \cos y) \, dy \, dx = 0$$

33. A possible parametrization is $\mathbf{r}(r,\theta) = r \cos \theta \, \mathbf{i} + r \sin \theta \, \mathbf{j} + (1 + r) \, \mathbf{k}$ (cylindrical coordinates) where $r = \sqrt{x^2 + y^2}$ $\Rightarrow z = 1 + r$. $1 \le z \le 3 \Rightarrow 1 \le 1 + r \le 3 \Rightarrow 0 \le r \le 2$. Also, $0 \le \theta \le 2\pi$.

37. Let $z = 1 - x - y \Rightarrow f_x(x,y) = -1$, $f_y(x,y) = -1 \Rightarrow \sqrt{f_x^2 + f_y^2 + 1} = \sqrt{3} \Rightarrow$ Area $= \iint_R \sqrt{3} \, dx \, dy =$ $\sqrt{3}$(Area of the circlular region in the xy–plane) $= \pi\sqrt{3}$

41. $\frac{x}{a} + \frac{y}{b} + \frac{z}{c} = 1 \Rightarrow$ x–intercept = a, y–intercept = b, z–intercept = c. $F = \frac{x}{a} + \frac{y}{b} + \frac{z}{c} \Rightarrow \nabla F = \frac{1}{a}\mathbf{i} + \frac{1}{b}\mathbf{j} + \frac{1}{c}\mathbf{k} \Rightarrow$

$|\nabla F| = \sqrt{\frac{1}{a^2} + \frac{1}{b^2} + \frac{1}{c^2}}$. $\mathbf{p} = \mathbf{k} \Rightarrow |\nabla F \cdot \mathbf{k}| = \frac{1}{c}$ since $c > 0$. Area $= \iint_R \frac{\sqrt{\frac{1}{a^2} + \frac{1}{b^2} + \frac{1}{c^2}}}{1/c} \, dA =$

$$c\sqrt{\frac{1}{a^2} + \frac{1}{b^2} + \frac{1}{c^2}} \iint_R dA = \frac{1}{2} abc\sqrt{\frac{1}{a^2} + \frac{1}{b^2} + \frac{1}{c^2}}$$

45. Because of symmetry $\bar{x} = \bar{y} = 0$. Let $F(x,y,z) = x^2 + y^2 + z^2 = 25 \Rightarrow \nabla F = 2x\,\mathbf{i} + 2y\,\mathbf{j} + 2z\,\mathbf{k} \Rightarrow$

$|\nabla F| = \sqrt{4x^2 + 4y^2 + 4z^2} = 10$, $\mathbf{p} = \mathbf{k} \Rightarrow |\nabla F \cdot \mathbf{p}| = 2z$ since $z \ge 0 \Rightarrow M = \iint_R \delta(x,y,z) \, d\sigma =$

$\iint_R z\left(\frac{10}{2z}\right) dA = \iint_R 5 \, dA = 5$(Area of the circular region) $= 80\pi$. $M_{xy} = \iint_R z\delta \, d\sigma =$

$$\iint_R 5z \, dA = \iint_R 5\sqrt{25 - x^2 - y^2} \, dx \, dy = \int_0^{2\pi} \int_0^4 5\sqrt{25 - r^2} \, r \, dr \, d\theta = \int_0^{2\pi} \frac{490}{3} \, d\theta = \frac{980}{3}\pi$$

$\therefore \bar{z} = \frac{\frac{980}{3}\pi}{80\pi} = \frac{49}{12}$. Thus $(\bar{x},\bar{y},\bar{z}) = \left(0,0,\frac{49}{12}\right)$. $I_z = \iint_R (x^2 + y^2)\delta \, d\sigma = \iint_R 5(x^2 + y^2) \, dx \, dy =$

$$\int_0^{2\pi} \int_0^4 5r^3 \, dr \, d\theta = \int_0^{2\pi} 320 \, d\theta = 640\pi. \quad R_z = \sqrt{I_z/M} = \sqrt{\frac{640\pi}{80\pi}} = 2\sqrt{2}$$

49. $\frac{\partial}{\partial x}(-2x) = -2, \frac{\partial}{\partial y}(-3y) = -3, \frac{\partial}{\partial z}(z) = 1 \Rightarrow \nabla \cdot \mathbf{F} = -4 \Rightarrow \text{Flux} = \int\int_D\int -4\, dV =$

$$-4\int_0^{2\pi}\int_0^{1}\int_{r^2}^{\sqrt{2-r^2}} dz\, r\, dr\, d\theta = -4\int_0^{2\pi}\int_0^{1}\left(r\sqrt{2-r^2} - r^3\right) dr\, d\theta = -4\int_0^{2\pi}\left(-\frac{7}{12} + \frac{2}{3}\sqrt{2}\right) d\theta =$$

$$\frac{2}{3}\pi\left(7 - 8\sqrt{2}\right)$$

53. $\nabla f = 2\,\mathbf{i} + 6\,\mathbf{j} - 3\,\mathbf{k} \Rightarrow \nabla \times \mathbf{F} = -2y\,\mathbf{k}.\ \mathbf{n} = \frac{2\,\mathbf{i} + 6\,\mathbf{j} - 3\,\mathbf{k}}{\sqrt{4+36+9}} = \frac{2\,\mathbf{i} + 6\,\mathbf{j} - 3\,\mathbf{k}}{7} \Rightarrow \nabla \times \mathbf{F}\cdot\mathbf{n} = \frac{6}{7}y.\ \mathbf{p} = \mathbf{k} \Rightarrow$

$$|\nabla f \cdot \mathbf{p}| = 3 \Rightarrow d\sigma = \frac{7}{3}dA \Rightarrow \oint_C \mathbf{F}\cdot d\mathbf{r} = \int_R\int \frac{6}{7}y\, d\sigma = \int_R\int \frac{6}{7}y\left(\frac{7}{3}dA\right) = \int_R\int 2y\, dx\, dy =$$

$$\int_0^{2\pi}\int_0^{1} 2r\sin\theta\ r\, dr\, d\theta = \int_0^{2\pi}\frac{2}{3}\sin\theta\, d\theta = 0$$

57. $dx = \cos 2t\, dt,\ dy = \cos t\, dt.$ $\text{Area} = \frac{1}{2}\oint_C x\,dy - y\,dx = \frac{1}{2}\int_0^{\pi}\left(\frac{1}{2}\sin 2t\cos t - \sin t\cos 2t\right) dt =$

$$\frac{1}{2}\int_0^{\pi}\left(-\sin t\cos^2 t + \sin t\right) dt = \frac{2}{3}$$

APPENDICES

APPENDIX A.2 PROOFS OF THE LIMIT THEOREMS IN CHAPTER 2

1. Let $\lim_{x\to c} f_1(x) = L_1$, $\lim_{x\to c} f_2(x) = L_2$, $\lim_{x\to c} f_3(x) = L_3$. Then $\lim_{x\to c}\left(f_1(x) + f_2(x)\right) = L_1 + L_2$ by Theorem 1. Thus $\lim_{x\to c}\left(f_1(x) + f_2(x) + f_3(x)\right) = \lim_{x\to c}\left(\left(f_1(x) + f_2(x)\right) + f_3(x)\right) = (L_1 + L_2) + L_3 = L_1 + L_2 + L_3$.

   Suppose functions $f_1(x), f_2(x), f_3(x), \ldots, f_n(x)$ have limits $L_1, L_2, L_3, \ldots, L_n$ as $x \to c$.
   Prove $\lim_{x\to c}\left(f_1(x) + f_2(x) + f_3(x) + \cdots + f_n(x)\right) = L_1 + L_2 + L_3 + \cdots + L_n$, n a positive integer.

   Step 1: For n = 1, we are given that $\lim_{x\to c} f_1(x) = L_1$.

   Step 2: Assume $\lim_{x\to c}\left(f_1(x) + f_2(x) + f_3(x) + \cdots + f_k(x)\right) = L_1 + L_2 + L_3 + \cdots + L_k$ for some k.
   Then $\lim_{x\to c}\left(f_1(x) + f_2(x) + f_3(x) + \cdots + f_k(x) + f_{k+1}(x)\right) =$
   $\lim_{x\to c}\left(\left(f_1(x) + f_2(x) + f_3(x) + \cdots + f_k(x)\right) + f_{k+1}(x)\right) =$
   $\left(L_1 + L_2 + L_3 + \cdots + L_k\right) + L_{k+1} = L_1 + L_2 + L_3 + \cdots + L_k + L_{k+1}$.

   $\therefore$ by Steps 1 and 2 and mathematical induction, $\lim_{x\to c}\left(f_1(x) + f_2(x) + f_3(x) + \cdots + f_n(x)\right)$
   $= L_1 + L_2 + L_3 + \cdots + L_n$

5. $\lim_{x\to c}\dfrac{f(x)}{g(x)} = \dfrac{\lim_{x\to c} f(x)}{\lim_{x\to c} g(x)}$ (by Theorem 1) $= \dfrac{f(c)}{g(c)}$ if $g(c) \neq 0$ (by Exercise 4).

APPENDIX A.4 MATHEMATICAL INDUCTION

1. Step 1: For n = 1, $|x_1| = |x_1| \le |x_1|$.
   Step 2: Assume $|x_1 + x_2 + \cdots + x_k| \le |x_1| + |x_2| + \cdots + |x_k|$ for some positive integer k.
   Then $|x_1 + x_2 + \cdots + x_k + x_{k+1}| = |(x_1 + x_2 + \cdots + x_k) + x_{k+1}|$
   $\le |x_1 + x_2 + \cdots + x_k| + |x_{k+1}|$ (by the triangle inequality)
   $\le |x_1| + |x_2| + \cdots + |x_k| + |x_{k+1}|$. $\therefore |x_1 + x_2 + \cdots + x_n| \le |x_1| + |x_2| + \cdots + |x_n|$ for all positive integers n by Steps 1 and 2 and mathematical induction.

5. Step 1: For n = 1, $\dfrac{2}{3^1} = \dfrac{2}{3} = 1 - \dfrac{1}{3^1}$.
   Step 2: Assume $\dfrac{2}{3^1} + \dfrac{2}{3^2} + \cdots + \dfrac{2}{3^k} = 1 - \dfrac{1}{3^k}$ for some positive integer k.
   Then $\dfrac{2}{3^1} + \dfrac{2}{3^2} + \cdots + \dfrac{2}{3^k} + \dfrac{2}{3^{k+1}} = 1 - \dfrac{1}{3^k} + \dfrac{2}{3^{k+1}} = 1 - \dfrac{3^{k+1}}{3^k 3^{k+1}} + \dfrac{2(3^k)}{3^k 3^{k+1}} = 1 - \left(\dfrac{3^{k+1} - 2(3^k)}{3^k 3^{k+1}}\right) =$
   $1 - \left(\dfrac{3^k(3-2)}{3^k 3^{k+1}}\right) = 1 - \dfrac{1}{3^{k+1}}$. $\therefore \dfrac{2}{3^1} + \dfrac{2}{3^2} + \cdots + \dfrac{2}{3^n} = 1 - \dfrac{1}{3^n}$ for all positive integers n by Steps 1 and 2 and mathematical induction.

9. Step 1: For $n = 1$, $1^2 = \frac{1(1 + 1)(2(1) + 1)}{6}$.

Step 2: Assume $1^2 + 2^2 + 3^2 + \cdots + k^2 = \frac{k(k + 1)(2k + 1)}{6}$ for some positive integer k.

Then $1^2 + 2^2 + 3^2 + \cdots + k^2 + (k + 1)^2 = \frac{k(k + 1)(2k + 1)}{6} + (k + 1)^2 \Rightarrow$

$1^2 + 2^2 + 3^2 + \cdots + k^2 + (k + 1)^2 = \frac{k(k + 1)(2k + 1) + 6(k + 1)^2}{6} \Rightarrow$

$1^2 + 2^2 + 3^2 + \cdots + k^2 + (k + 1)^2 = \frac{(k + 1)(k(2k + 1) + 6(k + 1))}{6} \Rightarrow$

$1^2 + 2^2 + 3^2 + \cdots + k^2 + (k + 1)^2 = \frac{(k + 1)(2k^2 + 7k + 6)}{6} \Rightarrow$

$1^2 + 2^2 + 3^2 + \cdots + k^2 + (k + 1)^2 = \frac{(k + 1)(k + 2)(2k + 3)}{6} \Rightarrow$

$1^2 + 2^2 + 3^2 + \cdots + k^2 + (k + 1)^2 = \frac{(k + 1)((k + 1) + 1)(2(k + 1) + 1)}{6}$.

$\therefore$ by Steps 1 and 2 and mathematical induction, $1^2 + 2^2 + 3^2 + \cdots + n^2 = \frac{n(n + 1)(2n + 1)}{6}\left(\frac{1/2}{1/2}\right) =$

$\frac{n\left(n + \frac{1}{2}\right)(n + 1)}{3}$ for all positive integers n.

APPENDIX A.7 DETERMINANTS AND CRAMER'S RULE

1. $\left|\begin{matrix} 2 & 3 & 1 \\ 4 & 5 & 2 \\ 1 & 2 & 3 \end{matrix}\right| \begin{matrix} 2 & 3 \\ 4 & 5 \\ 1 & 2 \end{matrix} = 30 + 6 + 8 - 5 - 8 - 36 = -5$

5. a) $\begin{vmatrix} 2 & -1 & 2 \\ 1 & 0 & 3 \\ 0 & 2 & 1 \end{vmatrix} = \begin{vmatrix} 2 & -5 & 2 \\ 1 & -6 & 3 \\ 0 & 0 & 1 \end{vmatrix} = 1\begin{vmatrix} 2 & -5 \\ 1 & -6 \end{vmatrix} = -7$

b) $\begin{vmatrix} 2 & -1 & 2 \\ 1 & 0 & 3 \\ 0 & 2 & 1 \end{vmatrix} = \begin{vmatrix} 2 & -1 & 2 \\ 1 & 0 & 3 \\ 4 & 0 & 5 \end{vmatrix} = -(-1)\begin{vmatrix} 1 & 3 \\ 4 & 5 \end{vmatrix} = -7$

9. $D = \begin{vmatrix} 1 & 8 \\ 3 & -1 \end{vmatrix} = -25$. $x = \frac{\begin{vmatrix} 4 & 8 \\ -13 & -1 \end{vmatrix}}{-25} = \frac{100}{-25} = -4$, $y = \frac{\begin{vmatrix} 1 & 4 \\ 3 & -13 \end{vmatrix}}{-25} = \frac{-25}{-25} = 1$

13. $D = \begin{vmatrix} 2 & 1 & -1 \\ 1 & -1 & 1 \\ 2 & 2 & 1 \end{vmatrix} = \begin{vmatrix} 2 & 1 & -1 \\ 3 & 0 & 0 \\ 4 & 3 & 0 \end{vmatrix} = -1\begin{vmatrix} 3 & 0 \\ 4 & 3 \end{vmatrix} = -9.$ $x = \dfrac{\begin{vmatrix} 2 & 1 & -1 \\ 7 & -1 & 1 \\ 4 & 2 & 1 \end{vmatrix}}{-9} = \dfrac{\begin{vmatrix} 2 & 1 & -1 \\ 9 & 0 & 0 \\ 6 & 3 & 0 \end{vmatrix}}{-9} =$

$\dfrac{-1\begin{vmatrix} 9 & 0 \\ 6 & 3 \end{vmatrix}}{-9} = 3,$ $y = \dfrac{\begin{vmatrix} 2 & 2 & -1 \\ 1 & 7 & 1 \\ 2 & 4 & 1 \end{vmatrix}}{-9} = \dfrac{\begin{vmatrix} 2 & 2 & -1 \\ 3 & 9 & 0 \\ 4 & 6 & 0 \end{vmatrix}}{-9} = \dfrac{-1\begin{vmatrix} 3 & 9 \\ 4 & 6 \end{vmatrix}}{-9} = -2,$ $z = \dfrac{\begin{vmatrix} 2 & 1 & 2 \\ 1 & -1 & 7 \\ 2 & 2 & 4 \end{vmatrix}}{-9} =$

$\dfrac{\begin{vmatrix} 3 & 0 & 9 \\ 1 & -1 & 7 \\ 4 & 0 & 18 \end{vmatrix}}{-9} = \dfrac{-1\begin{vmatrix} 3 & 9 \\ 4 & 18 \end{vmatrix}}{-9} = 2$

17. $D = \begin{vmatrix} 2 & h \\ 1 & 3 \end{vmatrix} = 6 - h = 0 \Rightarrow h = 6.$ x: $\begin{vmatrix} 8 & h \\ k & 3 \end{vmatrix} = 24 - hk = 24 - 6k = 0 \Rightarrow k = 4$

a) When $h = 6$, $k = 4$, there are infinitely many solutions.

b) When $h = 6$, $k \neq 4$, there are no solutions.